Hood River County Library

GIVEN IN MEMORY OF

GINA O'CONNOR

BY

HOOD RIVER COUNTY DEPARTMENT HEADS

The Emerald Sea

The Emerald Sea

Exploring the Underwater Wilderness of the
Pacific Northwest and Alaska

Photography by
Dale Sanders

Text by
Diane Swanson

Alaska Northwest Books™
Anchorage • Seattle • Portland

Photographs copyright © 1993 by Dale Sanders
Text copyright © 1993 by Diane Swanson

Published by arrangement with Whitecap Books Ltd., 1086 West Third Street, North Vancouver, British Columbia, Canada V7P 3J6

All rights reserved. No part of this book may be reproduced or transmitted in any form or by any means, electronic or mechanical, including photocopying, recording, or by any information storage and retrieval system, without written permission of the publisher.

Library of Congress Cataloging-in-Publication Data
Sanders, Dale, 1957-
   The emerald sea : exploring the underwater wilderness of the Pacific Northwest and Alaska / photographs by Dale Sanders ; text by Diane Swanson.
      p.   cm.
   Includes bibliographic references (p. 141) and index.
   ISBN 0-88240-450-4
   1. Marine biology—Northwest Coast of North America. 2. Natural history—Northwest Coast of North America. 3. Marine biology—Northwest Coast of North America—Pictorial works. 4. Natural History—Northwest Coast of North America—Pictorial Works.
I. Swanson, Diane, 1944-   . II. Title.
QH104.5.P32S26 1993
574.9'63—dc20   93-4966
                                                                                                     CIP

Edited by Elaine Jones
Cover design by Carolyn Deby
Interior design by Margaret Ng

All photographs by Dale Sanders except the following:
Gordie Cox, pp. 86, 116; Jeff Foott, p. 106; Ed Gifford, p. 18; Nick Lawlor, p. 98; Lou Lehman, p. 123; Neil McDaniel, p. 104-5; Gary McIntyre, p. 81; Bill Reed, p. 84.

Map by Phototype Composing Ltd., Victoria, B.C.
Typography by Computype, Vancouver B.C.
Color separations by Zenith Graphics, Vancouver, B.C.

Printed and bound in Canada by D. W. Friesen and Sons Ltd., Altona, Manitoba

Alaska Northwest Books™
An imprint of Graphic Arts Center Publishing Company
Editorial office: 2208 NW Market Street, Suite 300, Seattle, WA 98107
Catalog and order dept.: P.O. Box 10306, Portland, OR 97210
800-452-3032

Acknowledgements

To my parents Ron and Vera Sanders

The completion of the photography for this book was made possible in part by a research grant from The Royal Canadian Geographic Society. Many thanks to James W. Maxwell, Diana Roue, Ian Darragh, and everyone at the society for their invaluable support.

My thanks to Al Spilde and Dan Kannon, at Exta Sea Charters out of Nanaimo, for some spectacular days of diving all along the coast, on board Al's diveboat the *Sea Venturer*. Thanks also to John de Boeck, of *Clavella* Adventures out of Nanaimo, for showing me many great dives in the emerald sea.

If any book was ever made possible due to the help of friends, this is it. It was the wonderful times spent diving with these friends that really made this project worthwhile. Special thanks to Bill Reed; Gordie and Barb Cox; Nick and Val Lawlor; Gary McIntyre; Andrea Osborn; Randy Kashino; Dereck and Cathy Gale; Jim and Jeanie Cosgrove; Mark and Jan Yunker; Mike and Darlene Richmond of Abyssal Diving; Lou and Eleanore Crabe; Jim Borrowman and Bill McKay of Stubbs Island Charters; Eugene White; Dan Balla and Heidi Styacko; and Dave and Renate Christie of *Rendezvous* Dive Ventures.

My thanks to Andy Lamb for reviewing the captions, and to Diane Swanson for jumping on board as the writer. Diane's professionalism and patience was greatly appreciated, and the friendship of Diane and her family helped to make all the work enjoyable. Finally, the warmest thanks of all goes to my parents, for always being there for me, and for always believing in me.
—*Dale Sanders*

Special thanks go to Jim Cosgrove of the Royal British Columbia Museum for reviewing the text and to Timothy Swanson for sharing his enthusiasm for diving and his experience with life in the emerald sea.
—*Diane Swanson*

Contents

Preface	ix
Introduction	1
Chapter I: Where the Land Meets the Sea	5
Chapter II: Life Within the Sea	53
Chapter III: People and the Sea	113
Scientific Names of Species	139
Suggested Reading	141
Converting Metric Measurements	144
Index	145

Preface

Slipping beneath a green-gold canopy of kelp, I glide downwards along a steep rock ridge into the shadowy world below. On the edge of an undersea cliff, I stop and glance up to watch my diving companions as they descend towards me. Golden shafts of sun shimmer and dance with each passing wave. A shy harbour seal ghosts by us and then disappears. With each exhalation silvery bubbles cascade towards the surface, where they burst into the atmosphere from which they came. We drift deeper. Hanging motionless for a moment off a jagged wall of rock at a depth of twenty metres, I am suspended in a twilight world where all colours have been lost, except for the emerald green of the sea.

When I turn on my powerful underwater light, the colours suddenly return with startling brilliance. I am astonished at the incredibly rich and vivid abundance of life before me. Deep, velvety mounds of pink soft coral cover the wall in flowerlike splendour, while yellow encrusting sponges and anemones in nearly every colour of the rainbow fill any available space that the coral has missed.

Entranced, mesmerized by the scene before me, I watch a red Irish lord, as showy as any tropical fish, shuffle across the bottom on fanned-out pectoral fins. A small octopus ventures out of its den and moves off, gliding fluidly around the corals and sponges. An orange-peel nudibranch, laced with startling, pure white frills, crawls slowly along a mound of soft coral, feeding on the polyps as it goes. Gently I touch the coral and the polyps instantly retract, leaving the branch looking like it is decorated with wilted red strawberries. A current begins to tug me insistently along the reef. Our brief window of slack water is coming to an end, and soon the incoming tide will turn this small passage into a powerful river. I am not anxious to leave this bright coral garden, but I do not resent the currents. The nutrients and food they bring are responsible for this incredible bounty of life. Enjoying the free ride, we glide effortlessly along the wall, past a school of rockfish that hangs in the kelp forest above us. We linger for a short time in a quiet back eddy, then reluctantly begin our ascent to the surface.

* * *

Above: A red-gilled aeolid nudibranch in a purple colour phase passes several orange cup corals.

Previous page: A sunflower star patrols an offshore reef beneath the waters of the emerald sea. The largest sea star along the north Pacific coast, it can grow to over a metre in diameter. The number of arms increases with age.

We were diving a small, current-swept passage at the southern end of Queen Charlotte Strait, off northern Vancouver Island, and this exotic coral garden was without doubt one of the richest ocean realms that I had ever encountered. Until that dive, I had never known that anything like it existed off this coast.

After diving and photographing marine life for several years in the Caribbean, I discovered that one of the most spectacular underwater wilderness areas in the world had existed all along virtually in my own backyard—the rich coastal waters that extend from Oregon and Washington, along British Columbia, to Alaska. Even though I had enjoyed hundreds of dives over the years in these cold green waters, I still knew very little of what was out there. It was on this dive that the idea for *The Emerald Sea* was born.

Since that day I have had many other special dives and experiences while exploring this vibrant ocean realm. I have swum with dolphins and sea lions, and watched in awe as a massive grey whale passed slowly by me in only three metres of water. I have explored shipwrecks, ranging from the rusted, anemone-covered hulks of old steamships to the massive 112-metre-long HMCS *Chaudiere,* a former Canadian naval destroyer escort sunk in 1992 as part of an artificial reef program. I have drifted over magical reefs adorned with pink and white gorgonians, a tropical-looking coral whose existence in these waters was not even known until a few years ago.

Many other discoveries have been made by divers in recent years, and many more have yet to be made. That is one of the wonderful things about diving in the emerald sea. With its numerous islands, passages, and inlets, well over fifty thousand kilometres of coastline are waiting to be explored. Each type of marine habitat offers a very different undersea realm—ranging from deep mountain fiords to an island-studded inland sea, and from a wild, surf-swept, outer coast to inside passages sporting the fastest tidal currents recorded anywhere in the world.

It is impossible to photograph the incredible natural wonders of this underwater wilderness without developing some concern for its future. The massive oil spill created when the supertanker *Exxon Valdez* ran aground in Prince William Sound in 1989 opened many eyes to the fragility of the marine ecosystems of these northern seas. For far too long we have looked upon the ocean as an endless source of food and a bottomless cesspool for our wastes. But if our emerald sea is to survive in its relatively pristine state, it is going to need our protection.

On land we try to protect great natural wilderness areas by creating parks and sanctuaries. Unfortunately, we haven't yet extended that same attitude to the wilderness hidden beneath the sea. Those marine parks we do have are often simply places to anchor a boat, created to protect their above-water environments, not the life below the waves.

It is my hope that this book will help people see the wonders that lie within the emerald sea, so that they may help to preserve it for future generations. I would like to return someday to that current-swept passage in Queen Charlotte Strait and know that all life that dwells there is protected. For, although this book is now complete, I know that my exploration of this fascinating sea has just begun.

—*Dale Sanders*

Overleaf: Raspberrylike hydroids, of a species not yet scientifically described by biologists, compete for space with a giant barnacle and plumose anemones. These plantlike, primitive animals use stinging cells to capture prey.

Below: A small octopus roams through a garden of pink soft corals, anemones, and sulphur sponge.

Introduction

Stretching along the coasts of British Columbia, Washington, Oregon, and Southeast Alaska is a sea so thick with microscopic plants that its waters are emerald green. This abundance of plant life, the basis of the entire food chain, supports vast numbers and many species of animals, making the "emerald sea" one of the most productive marine environments anywhere in the world.

Variations in topography, currents, upwellings, and wave action within the emerald sea have created differing underwater habitats that support a wide range of both plants and animals. Steep-walled mountain fiords cradle still waters, forming quiet realms for fluffy cloud sponges and wispy feather stars. Sheltered inner coasts rim active, but protected, waters that harbour wintering sea lions and large schools of herring. Along exposed outer coasts, wind-beaten waters surge over dense kelp stands, tightly grasping sea palms, and giant mussels. And narrow tidal passages, lined with soft coral and carpets of sea anemones, channel some of the planet's most turbulent waters.

As enticing as this underwater world is, relatively few people visit it. For most, images of the emerald sea are defined by the camera lenses of scuba divers who explore the upper thirty metres of the sea, recording its fascinating life: seaweed growing in forests as tall as multistorey buildings; invertebrates, such as octopuses and sea stars, moving their spineless bodies across the sea floor; fish hovering inside ancient shipwrecks and peering out from giant barnacle shells; and marine mammals—the warm-blooded creatures of this cold sea—responding with curiosity to human intruders.

Although the life in this sea seems boundless, the last few centuries have seen a drastic reduction—even extinction—of some species. Gone are large populations of such magnificent creatures as sea otters, humpback whales, and plankton-feeding basking sharks. Gone are unknown numbers of humble creatures, small invertebrates whose populations declined or disappeared before scientists understood the roles they played in the ecosystem. These losses, often the results of overharvesting and pollution, are the sea's unfortunate legacy of interaction with people.

Still, even among those who have never entered the underwater realm, there is a growing awareness of the sea, an appreciation for its beauty, and a respect for its needs. There is also an abiding fascination with its mystery. For in its vastness, the emerald sea holds enormous potential for discovery—for finding life in new dimensions or new places. That potential has barely been tapped.

Previous page: The young of brooding anemones emerge from the mouth of the parent anemone, then slide down the column and attach themselves to the base until they are big enough to move off on their own.

Above: A rock painted with purple encrusting hydrocoral contrasts colourfully with a blood star. Brightly coloured hydrocorals grow abundantly where there is strong tidal movement.

Overleaf: A vermilion star and giant red sea urchins add colour to the shallows on the protected inner coast.

Chapter I
Where the Land Meets the Sea

*"...no boundaries, no beginning, no end,
one continual shove of growing—edge of land
meeting edge of water..."*
—*West Coast artist Emily Carr*

Rocky headlands, raging waves, barrier islands, still waters. The long, convoluted coastline of the emerald sea is a product of the constant interaction between land and ocean. Along Oregon and most of Washington, waves spend themselves against a relatively even coast, but from Puget Sound, along British Columbia to Alaska, land and water entwine. Deeply

Right: The elegant red and white plumes of a calcareous tube worm, also known as a plume worm.

Previous page: A crimson anemone's stinging cells are concentrated in the darker bands that pattern its tentacles.

probing fiords, low undercut bluffs, and intimate groupings of islands reveal "no boundaries, no beginning, no end," as Emily Carr aptly described the coast. So interconnected are land and sea that, for every kilometre the crow flies, there are roughly thirty kilometres of curving, twisting shoreline.

This intimacy between land and sea does not end at the shore: it continues beneath the waves, shaping the various underwater environments that support the abundant marine life characteristic of this sea. Nor is the intimacy static. The contest between land and sea is played out continuously, as it has been for centuries.

About 200 million years ago, the southwestward-flowing continent of North America slowly, but forcefully, met the northward-moving floor of the Pacific Ocean. The prolonged impact, which continues even today, caused the heavy sea floor to sink gradually into the planet's hot interior and rugged mountains to rise steeply along the continent's edge.

6 THE EMERALD SEA

Unlike the much smoother Atlantic coast, the Pacific coast is craggy and abruptly sloping. At 6,194 metres, Denali (Mount McKinley), in the coastal Alaska Range, is the tallest mountain in North America. Canada's tallest, Mount Logan (5,950 metres), rises right from the sea among the jagged St. Elias Mountains of Yukon Territory.

About 150 million years ago, another major coastal feature began to form—a great trench, or depression, that now stretches four thousand kilometres from California to Alaska. Beginning in the Gulf of California, the Coastal Trough passes through California, Oregon, and Washington above sea level, then submerges at Puget Sound. It continues along British Columbia as the Strait of Georgia, Queen Charlotte Sound, and Hecate Strait, wedged between the mountains on the mainland coast and those on Vancouver Island and the Queen Charlotte Islands. Extending through the Alaska Panhandle, the Coastal Trough finally ends in Chatham Strait, Alaska.

This protracted trough and the great mountains of the coast and islands continued to undergo a series of gradual uplifts and downturns. Then great sheets of ice scoured the region during several periods of glaciation, dramatically sculpting scores of steep-walled fiords out of much of the coastline. As the ice receded, water drowned parts of the shore.

The submerged portions of this shore form part of North America's western continental shelf. Created by combinations of glaciation, sinking land, silting, and wave action, the shelf creates

A boldly coloured tiger rockfish. Rockfish are opportunistic predators that feed on prey such as small fish, crustaceans, and worms. Their dorsal fins have twelve to fifteen slightly venomous spines.

Right: A moon jellyfish pulses gently through the waters of a coastal inlet. These ethereal creatures are most abundant during the summer months.

Opposite: On a sunny winter day, a diver floats above a dense cluster of purple, or ochre, sea stars in the clear waters of a coastal inlet. Generally purple in colour, these sea stars are often orange on the outer coasts.

a relatively shallow sea floor that skirts the coast north to south. At the shelf's edge, the floor slopes—often steeply—down to the deep ocean bottom. Although some stretches of the shelf are topographically bland, most feature the same rugged characteristics as the shore above them, modified only by erosion and deposits of sediment.

Generally, this continental shelf is much narrower than its Atlantic counterpart, but the width varies considerably from one location to another. Off the south end of Vancouver Island, the biggest island on North America's Pacific coast, the shelf spreads westward for about ninety kilometres, but off the west coast of the Queen Charlotte Islands, it is so narrow as to be non-existent.

Water temperature is critical to distribution of sea life along this shelf, and several factors influence that temperature. The Kuroshio, or Japanese, Current, which moves off the shores of Japan and continues across the Pacific Ocean as the North Pacific Drift, branches north and south as it approaches North America, pushing cold waters along the coast. Strong winds and the rotation of Earth combine to move surface waters that are above the

A quillback rockfish, one of more than thirty species of rockfish in the emerald sea, rests on a plush carpet of strawberry anemones.

shelf out to sea, while cooler, deep-ocean waters from one hundred to three hundred metres below slowly well up to replace them. These waters mix with the current-pushed waters, creating a cold coastal sea (averaging ten degrees Celsius in surface waters) that extends from Oregon to Southeast Alaska.

These upwellings of water from the deep ocean raise phosphates, nitrates, and other nutrients from the darkness where there is no plant life to deplete them. At the surface, they mix with more nutrients brought to the sea by runoff from rivers. So well-nourished are the waters above the continental shelf that they support an abundance of microscopic green algae, called phytoplankton. It grows in sunlit waters, using energy produced through photosynthesis. The ability of cold water to hold large amounts of carbon needed for photosynthesis also promotes the growth of phytoplankton. Because this phytoplankton grows so profusely above the continental shelf, most other forms of sea life in the food chain also gather there.

Each year, however, unpredictable currents deliver projections of blue oceanic water to this cold sea, bringing with them a variety of sea life that is not common here. Off the Queen Charlotte Islands, for instance, there have been sightings of huge ocean sunfish, as well as soupfin sharks, white sharks, masses of blue sharks—even giant sea turtles.

Sporadically, El Nino, the warm Peruvian current that has sometimes preceded devastating global events, further alters the distribution of life in these waters, even close to shore. During

1982, warm waters from El Nino affected populations of sea life as far north as Southeast Alaska. Temperatures rose by two to four degrees Celsius, driving out cold-water fish, such as salmon, and drawing in warm-water fish, such as barracuda and bonito.

Although temperature and nutrient supplies are critical to the distribution and abundance of most species of marine life, so is topography. Life on the Pacific shelf partly reflects its diversity: rocky to sandy, and steep to shallow—and the fact that it has few wide-open spaces. This shelf supports a much wider range of life than its counterpart on the Atlantic, which is more topographically uniform.

Currents and wave action also affect the distribution of phytoplankton and the type of life that feeds on it. Within the emerald sea, there are four general habitats, roughly characterized by the movement of water within them: calm-water fiords, which typically experience low rates of circulation; sheltered waters along protected inner coasts, where movement is stronger; exposed waters off outer coasts, which experience powerful waves and ocean currents; and turbulent tidal passages, where currents reach record-breaking speeds. Although there is considerable overlap, each of these habitats supports varying types of sea life in varying degrees of abundance; each has its own attractions—its unique appeal.

Fiords: Calm Waters

Guarding the narrow entrance to a typical west coast fiord, stone-faced giants stand straight and tall, their steep slopes scarred by glaciers and gashed by landslides. The waterway stretches far inland, twisting and turning among snow-capped

Typical of many coastal fiords, Jervis Inlet stretches more than eighty kilometres into the coastal mountains and reaches a maximum depth of over 730 metres.

A crinoid, or feather star, is able to swim to escape danger by gently beating its featherlike arms in unison.

mountains where icy streams cascade hundreds of metres down rocky walls.

A fiord commonly funnels and compresses thick coastal clouds, then thrusts them upwards with a suddenness that forces release of their moisture. Rain pours down in torrents, pocking the surface of portions of the waterway. Viewed from beneath the water, the scene is striking. Divers look up to a hypnotic maze of overlapping circles that frame raindrops when they hit. Then, when sunshine manages to burst through the clouds, hazy rays play across the surface, introducing a dramatic change of scene.

Diving in a fiord is an exercise in tranquillity. It is a world where delicately fringed jellyfish journey slowly up and down, their bodies throbbing rhythmically as they go. With the winds and gently flowing water, they often drift along, moving slowly to the head and down the arms of the fiord. There they mass—sometimes so thickly that jellyfish are all a diver can see in any direction.

Cloud Sponge

Drifting out over the immense rock faces of the undersea mountain that forms the wall of the fiord, I feel tiny and insignificant. The water is clear as glass on this sunny winter day. Glancing upwards from thirty metres below the surface, I can still see the boat, and even the tree that it is tied to on shore. Below me, billowy white clouds appear to be floating above the dark rock. They are giant cloud sponges, which cover much of the wall for as far as the eye can see. I photograph my diving companion as she moves in close to examine one of the sponge colonies. Looking a little like a distorted clump of brain coral, parts of it extend outwards, resembling a collection of horns and trumpets. Pointing out a small rockfish sheltered inside the long-living sponge, she moves around the fragile colony with extreme care. One errant kick of a diver's fin can shatter a huge sponge colony that may have taken a hundred years to grow.

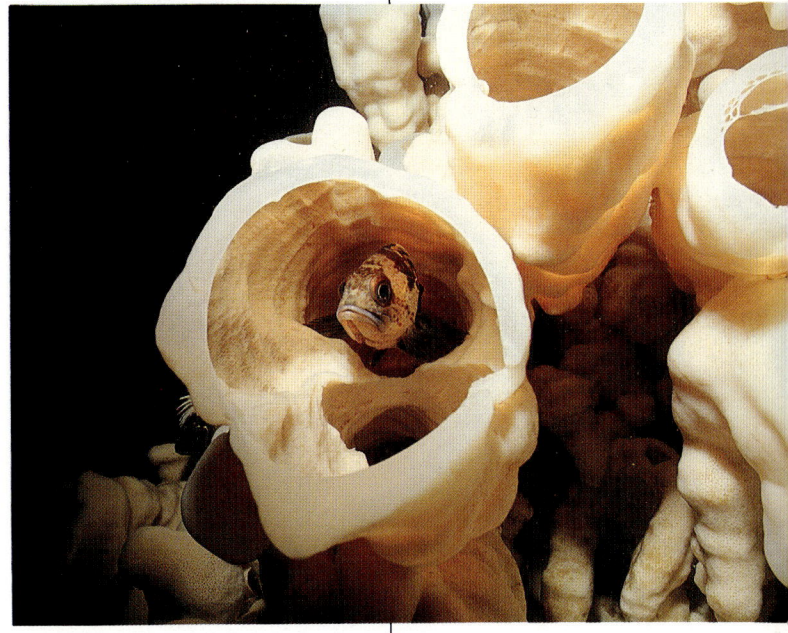

During winter, the water may be so clear that it is possible to look up from depths of thirty metres to see the branches of overhanging trees and the mountain slopes above. But in the summer, the upper waters of fiords may be thick with clouds of plankton. Divers pass through these clouds, seeing nothing until they get beneath the mass. There, the water may be perfectly clear, but the thick plankton absorbs and reflects so much of the sunshine that divers need artificial light to see anything at all.

Even when the sunlight can penetrate, most divers use artificial light to restore the breathtaking colours within the sea. Like a filter, water absorbs certain colours from natural light, and this absorption increases with depth. Red light, for instance, disappears completely—even in clear, fairly shallow water—making red objects appear a dull brown, grey, or black.

Fiords tangle much of the mainland coasts of British Columbia and Southeast Alaska as well as the west coasts of Vancouver Island, the Queen Charlotte Islands, and the islands of the Alaska Panhandle. Although many of these fiords are not known worldwide, they rank in size and splendour with their famous counterparts in Norway and New Zealand.

Deep-water Gorgonian

Peering even deeper, we are irresistibly drawn by a sight I have never seen before. A rarely seen deep-water gorgonian coral, more than a metre high, spreads its exotic fans in the perpetual twilight of the depths. Illuminated by our dive lights, its brilliant red colour seems oddly out of place in this world without sun. Its delicate polyps reach out with tiny tentacles to feed on plankton that drifts down from the sunlit waters above. Glancing at my decompression meter, I see that we are now at a depth of fifty metres—we are deeper than we should be, so after a brief glimpse we ascend to the safety and comfort of the shallows.

Reaching lengths of up to four hundred kilometres and widths up to fifteen kilometres, their waters usually run deep, walled by steep mountain slopes that plunge well below the surface. Parts of Finlayson Channel in British Columbia reach more than 750 metres in depth and the south end of Chatham Strait, Alaska, is at least 1,000 metres beneath the surface. Many fiords are navigable by large ships—even close along the edges.

Most fiords started out as V-shaped river valleys. Over millions of years, gouging ice enlarged and transformed them into U-shaped trenches. The enormous weight of glaciers caused the land to sag; then, as the ice melted, the sea flooded the scoured trenches and created fiords. Fine rock, which was concentrated in the melting ice, turned the water milky white before settling, a process that continues today. It is this silt that has helped to create the flat bottoms so characteristic of fiords.

Some icefields are still retreating, and some of them quite quickly. Those in Glacier Bay, Alaska, for example, have withdrawn more than one hundred kilometres during the last two hundred years, producing a series of dramatically beautiful fiords.

Receding glaciers typically create shallow entrances to fiords by depositing boulders, crushed rock, silt, and clay in underwater ridges called sills. The tops of these sills reach from within a few metres to five hundred metres of the water's surface, often marking a pronounced change in water depth from one side to the other. In Jervis Inlet, British Columbia, the entrance drops suddenly from a depth of three hundred metres at the sill to six hundred metres just inside it.

Stacked high with glacial debris and uplifted by the rebounding Earth, the occasional sill rises above sea level and divides the fiord from the ocean. One example is British Columbia's Powell Lake, the bottom half of which is salt water. Several thousand years ago, Powell Lake was a fiord connected to the Strait of Georgia.

Sills not only protect fiords from strong ocean currents and waves, they contribute to the low rate of circulation typical in these calm waterways. Glacier- and rain-fed rivers regularly contribute supplies of fresh water, which form a top layer a few metres thick. As it flows seaward, this water draws some of the fiord's salt water along with it. The sea then flows in to replace the fresh water. To the extent that a sill projects into the layer of outflowing water, it affects the degree of circulation within the fiord.

Crinoids, an ancient form of life related to sea stars, cling to a small white sponge in a coastal inlet.

The Shallows

Back in the comfort of the shallows after an enchanting dive in the depths of a fiord, I feel a warm contentment deep within myself. I take the time to examine the small creatures that are often overlooked. A delicate opalescent nudibranch, only three centimetres long, crawls slowly over an algae-encrusted rock. A small decorator crab hidden in a crack in the rocks appears to be grooming the garden of seaweeds and sponges that grows on its back. A tangle of purple sea stars completely covers a small section of reef. Above me, a snorkeler is briefly silhouetted against a backdrop of sky and trees.

Life in these waters tends to be limited in variety, but where nutrient and oxygen levels are high, some species of sea life flourish. Sea stars, for instance, cluster so thickly that they cloak entire walls of rock in purples and oranges. Eelgrass, a type of false grass, grows well at fiord heads in waters richly nourished by rivers; in turn, it supports communities of fish, such as seaperch.

In the shallows of many fiords, there is often a plentiful growth of kelp, a brown seaweed that sustains large numbers of small animals, such as crabs and beautiful, turban-shaped top snails. But deeper down, the amount of life usually decreases; walls grow barren and thin layers of silt carpet the slopes.

At depths of twenty to twenty-four metres, the silt dirties big boot sponges that grow up to a metre long off fiord walls. The insides of these stationary, tube-shaped animals, however, manage to remain clear of silt and provide shelter for crabs and rockfish.

Also at these depths, feather stars, ancient relatives of the common sea star, can carpet slopes as far as the eye can see. Waving their long arms up and down, they sometimes move short distances through the water. With the arrival of a voracious sunflower star, the feather stars may swim en masse, creating a thick sea of yellowish feathers. Young stars, which grow on stalks attached to rocks, look more like underwater lilies.

One of the giants of the emerald sea, the cloud sponge, often lives at depths of twenty-seven to thirty metres. Besides being abundant, it shares two other features with species that are characteristic of fiord life: a colossal size and a tendency to dominate the immediate landscape. Measuring up to two metres in diameter, this massive sponge resembles a fluffy, trumpet-shaped cloud. It grows for possibly one hundred years in a thick mass with many other cloud sponges, all attached to rock walls. Blending with the darkness of the sea at that depth, the walls are barely perceptible, creating the illusion

Opposite: A diver approaches several large colonies of cloud sponges that cling to the steep wall of an inlet. Some biologists speculate that colonies such as these may be over one hundred years old.

that the white and yellowish sponges are floating like clouds in a dark sky. Brown and yellow quillback rockfish often take shelter inside them.

Deeper still, in a few isolated fiords, live the most stunning creatures ever found in these calm waters: colourful colonies of giant gorgonian coral that grow more than a metre tall. Forming graceful fans of brilliant reds and oranges, this coral is an outstanding exception to fiord life, most of which lacks much colour. However, as it usually lives below the range of sport diving, the gorgonian coral remains one of the emerald sea's elusive treasures.

Inner Coasts: Sheltered Waters

Cutting across the gently rolling water off southern Vancouver Island, a small motorboat makes a beeline for a tiny, nondescript rocky island. As the current grows slack, divers plunge eagerly off the boat to explore the island's underwater base, which is as rich with life as its surface is barren.

The shallow waters surrounding the island are thick with the greens of sea lettuce and the browns of several species of kelp—all material for decorator crabs, which deliberately stick bits of sea life on themselves as camouflage. Below that, the sea has undercut the island, creating an overhang draped with tall, white plumose anemones and a steep rock wall graced with yellow staghorn bryozoans (colonies of microscopic animals that resemble coral). Narrow crevices here and there house a copper rockfish, a solitary octopus, and a peculiar-looking fish called a mosshead warbonnet. On the bottom, a massive Puget Sound king crab sits like a tank among the rocks, but in sandy stretches, feathery sea pens wave in the gentle currents. As the divers explore boulders, caves, and arches of rock, schools of black

A ferry cruises through British Columbia's Gulf Islands, with the Strait of Georgia and the coastal mountains in the background.

18 THE EMERALD SEA

A masking crab displays an exotic garden of plants, sponges, and other animals on its back. The growth helps camouflage this slow-moving crab from predators.

rockfish occasionally surround them, and now and then, a fast-diving sea lion pops into view.

Sites such as this one are common among inner coast waters, which are shielded year-round from the open ocean by large islands off the North American mainland. The Alaska Panhandle contains many of these sheltered inner coasts, but the best-known examples sit at the opposite end of the Inside Passage, in the Strait of Georgia-Puget Sound.

Protected from the Pacific Ocean by Vancouver Island, the Strait of Georgia is about 230 kilometres long. It links with waters off exposed coasts through Johnstone Strait and Queen Charlotte Strait on the north and through Juan de Fuca Strait on the south. Glacial erosion of Vancouver Island created the Gulf Islands—more than two hundred islands, islets, and reefs that break up the strait.

This archipelago continues across the Canada-United States border as the San Juan Islands, a group of more than 450 islands, islets, and reefs in the two-hundred-kilometre-long Puget Sound region. Bounded by the Olympic Peninsula on the west and the mainland on the east, the sound probes southward as far as Olympia, Washington, and links with waters off exposed shores through Juan de Fuca Strait.

Each time the tide changes, water is exchanged through these linking waterways. In some parts of the Strait of Georgia-Puget Sound, the mixing of the water is fast and complete, but in bays and other areas farther removed from the Queen Charlotte, Johnstone, and Juan de Fuca straits, the exchange is less.

Protected Seas

Sunrays dance across the bottom of a shallow reef, revealing the movement of fish and crabs seeking shelter beneath seaweed. A red sea star and a scattering of red sea urchins add a splash of colour to the tranquil scene. In an undersea canyon adorned with stately columns of white plumose anemones we find ourselves surrounded by a small school of seaperch. Our lights illuminate several bright crimson anemones crowning a rocky outcrop, and moving in close, we can see several brilliantly coloured shrimp moving around the bases of the anemones. The bodies of the shrimp match the anemones' colour perfectly, except for several fluorescent yellow and blue stripes that seem to glow with a light of their own. These beautiful clown, or candy stripe, shrimp commonly live with crimson anemone, seeking shelter and probably leftover morsels of food among the anemones' stinging tentacles. Gently coaxing a clown shrimp up the stalk of an anemone, I photograph it against the wonderfully textured backdrop.

Runoff from the rivers that drain into the Strait of Georgia-Puget Sound also affects its waters, decreasing the salinity and increasing the amount of sediment. Several small rivers from the Cascade Mountains on the mainland flow into Puget Sound, while about 75 per cent of fresh water reaching the Strait of Georgia comes from the mighty Fraser River. From late spring to early summer, the rivers spread a layer of light-coloured, silt-laden water up to ten metres thick over a sizable portion of this inland sea.

River runoffs combined with exchanges between the ocean and the sheltered waters contribute to a flushing action in the Strait of Georgia-Puget Sound that helps maintain its quality. Existing water is replaced by water that is rich in oxygen about once a month in the top thirty metres and about once a year at deeper levels.

Waves that cross the sheltered waters of this basin tend to have relatively low energy. The Gulf and San Juan Islands interrupt the movement of wind waves and hamper their growth. However, movement in the basin is highly variable from one portion to another and the Strait of Georgia is sufficiently open in stretches to allow development of moderate waves. Where wind waves confront water currents, higher waves occur. As strong currents stream out of Discovery Passage between Quadra and Vancouver islands, for example, they meet waves whipped up by southeast winds. The clash builds waves that are tall and steep enough to be potentially destructive to small boats.

Most of the populations of British Columbia and Washington live in the Strait of Georgia-Puget Sound area, which accounts

Opposite: Kelp greenlings swim placidly above a featherlike orange sea pen. Related to the soft corals, the sea pen may grow to a metre in height. Numerous polyps along the "branches" of its plumes capture small prey with tiny tentacles.

A dogfish cruises above a sandy bottom. The most commonly seen shark in these waters, dogfish are sometimes observed travelling in large schools.

for about 5 million people—1.5 million concentrated in Vancouver and 2 million in Seattle. Because the area is so highly populated, the basin forms a major marine playground and resource centre, offering some of the most accessible diving sites anywhere.

At Edmonds, Washington, for instance, a grid of pilings at the end of a pier extending 460 metres into Puget Sound has become a special habitat for a wide range of sea life. In the plankton-rich waters of this popular dive spot, masses of feathery plumose anemones and giant tube worms filter food from the water, while striped seaperch wind among them. Many other animals crowd the pilings for space: sponges, barnacles, corallike bryozoans, nudibranchs, and crabs. On the inner side of the pier, a sandy slope supports crowds of orange sea pens sixty centimetres tall, their feathery "quills" turned at right angles to the currents for optimum feeding. Small nudibranchs live in the sand, where they can prey on the sea pens.

Even well within cities, divers find good places to drop into the water. The Ogden Point Breakwater, in a harbour near downtown Victoria, British Columbia, is thought to be one of the most frequently dived sites in Canada. The kilometre-long, granite-block wall and its immediate waters shelter nearly every kind of life along the inner coasts: sponges, chitons, sea urchins, kelp, crabs, perch, scallops, barnacles, abalone, sea cucumbers, sea anemones—even wolf-eels.

Killer whales, or orcas, also frequent waters close to populated centres in the Strait of Georgia-Puget Sound. They draw large crowds of people to shores and to the decks of ferries that run between the islands and the mainland. Few sights are as inspiring as several of these sleek, black and white whales leaping in and out of the water in unison, their tall dorsal fins standing like sails on their backs. One community of eighty-nine killer whales that is often spotted during the summer are residents of southern sections of the Strait of Georgia-Puget Sound as well as parts of the outer coast.

Other familiar residents of bays and inlets along inner coasts are harbour seals. They frequently appear at fishing docks, looking for scraps, and divers often encounter them in golden-brown forests of bull kelp.

During the winter, rowdy Steller's and California sea lions leave rough waters off exposed outer shores and take up residence in the sheltered Strait of Georgia-Puget Sound. They haul out onto rocky islets and islands—even onto log booms—and take over for the season. During their stay, their constant barking shatters the air in every direction for several kilometres. Divers even hear them in the water as the sea lions swoop curiously around them.

A late winter treat for sea lions is a feed of Pacific herring, one of the sea's most abundant fish. So thickly do they mass in the shallows to spawn that they often turn the water white with a combination of the milt they release and the silt they stir up. The appearance of the herring precipitates a feeding frenzy. Sea lions charge after the fish, even chasing them right to shore, and as seabirds join the feast, the ruckus rises.

When herring are not spawning, they often travel in highly coordinated schools, forming a thick, silvery mass that moves

Overleaf: An opalescent nudibranch glides over a rock encrusted with coralline algae. The colour of the opalescent nudibranch varies widely, but it can easily be identified by the orange stripe that runs through the middle of its head.

Below: The sharp spines of a giant red sea urchin discourage most predators. These sea urchins, which graze on kelp and other algae, grow up to twenty centimetres in diameter.

A diver encounters a giant Pacific octopus. The largest known species of octopus in the world, they can grow to seventy kilograms or more.

in unison, opening and closing like parts of a single organism. Schools of spiny dogfish sharks often circle the herring, while auklets soar underwater to snatch some fish. Even a nine-metre minke whale will sometimes lunge through the mass of herring for a meal.

Another of the sea's abundant fish, the Pacific salmon, is hunted by dogfish sharks, sea lions, and whales. Salmon spend much of their adult lives in the emerald sea before readjusting to fresh water and journeying upstream to spawn. The Fraser River sockeye salmon run is reputed to be the largest in the world.

The Gulf and San Juan Islands offer a smorgasbord of some

Sea Lions

Suddenly a huge shape flashes by me, so close it blocks out my view of the reef: we have been joined by several large Steller's sea lions. Soon we are surrounded by these graceful clowns of the sea, which dive and pirouette around us. When I have almost run out of film, I lie on the bottom to watch them. One of them plops down in front of me and ambles over to check out my camera. My last photograph is a shot of a sea lion's nose and whiskers. Engrossed in my new companion, at first I don't even notice when several more curious sea lions join us. I look up to find I am encircled by a growing pyramid, as even more sea lions park themselves on the backs of their companions. I can hear the muffled laughter of my diving partner somewhere nearby. I hate it when I run out of film.

of the region's most exciting dive sites, including dramatic underwater cliffs, life-encrusted reefs, and shipwrecks. The area is habitat for a wide range of sea life, and the influx of nutrient-bearing water through tidal passages and waterways linked with the outer coast supports an abundance of species.

The giant Pacific octopus often lives here, in dens among rocks and reefs. Far from being the fearsome sea monster some people believe it to be, the octopus is usually reclusive, preferring to avoid contact with human beings.

Other common reef dwellers include a variety of attractive species of fish, such as kelp greenlings. In the Strait of Georgia-Puget Sound, where many divers feed them, greenlings are among the friendliest of fish, often easily tamed. Tiny, striped grunt sculpins, which usually vibrate or "grunt" when handled, regularly hide in crevices among rocks.

Away from the constricted passages of the Gulf and San Juan Islands, the more open stretches of water in the Strait of Georgia-Puget Sound generally experience little circulation. Surface currents often move through their wave-swept upper levels, supporting mussels, barnacles, crabs, abalone, sea anemones, sea urchins, rockfish, and bull kelp.

But at lower levels, where there is little water movement, the sea life is similar to that in fiords; jellyfish, quillback rockfish, and cloud sponges are common. Along rocky walls lives one of

the most striking invertebrates—the candy stripe, or clown, shrimp—whose transparent body is decorated in blazing reds, yellows, and neon blues. With a few others of its species, it usually inhabits the columns of graceful crimson anemones, where it can thrive on falling food particles and elude predators that stay well clear of the anemone's stinging tentacles. Along the muddy and sandy bottom, the giant nudibranch, which is one of the largest slugs in the emerald sea, reaches lengths of up to twenty-five centimetres. Usually softly coloured a greyish white and dabbed with purple, the shaggy, rather ethereal, form of this giant seldom fails to impress divers.

Above: A hermit crab carries its home on its back—an abandoned oregon triton shell.

Opposite: Anemones and sea stars crowd the surge-swept shallows of a west coast reef.

Outer Coasts: Exposed Waters

Born of strong winds far offshore, powerful Pacific waves travel unceasingly across thousands of kilometres of open sea. Raging storms in their path amplify their enormous store of energy. Then, as they reach the continental shelf and begin to touch bottom, the waves reduce their speed and rise up. Their crests grow steeper and steeper until they surge to the shore as breakers.

Unimpeded by islands or major reefs, the waves strike exposed stretches of the outer coast with extraordinary force. These stretches regularly endure breakers of enormous size and power—some of the heaviest surf anywhere in the world. Waves twenty metres high are not uncommon, especially between

Large Pacific breakers crash heavily on exposed shores. The stronger the wind and the longer the expanse of open ocean it crosses, the higher the waves. Inhabitants of the shallows of the outer coast have adapted to the harsh conditions found there.

November and February. Some display exceptional force: records show one wave off Oregon tossed a sixty-kilogram rock forty metres into the air to a lightkeeper's roof.

The combined forces of mighty waves, strong ocean currents, and savage winds exert hundreds of kilograms of pressure on coasts, carving sizable sea caves in places. Years of unrelenting beatings cause some of the shores to retreat through erosion or break off as islands or arches of rock. Other shores expand, as the steady waves build up beaches of coarsely ground pebbles or finely ground sand. Sometimes these shores grow to join islands or rocks offshore, scoring a battle victory for the land mass.

Trees that dare to encroach upon the edges of these wild shores—or the cliffs above them—often grow slowly, their twisted trunks and branches a testament to the harsh growing conditions. Shrubs, too, are wind-pruned and frequently scraggly, but on many stretches of rainy open coast, banks of shiny green salal persist, tangled and impenetrable.

Diving in these exposed waters calls for practised technique and steady confidence, especially along rough, rocky shores. The breaking waves make sea entries and exits difficult. In shallow water, where the waves touch bottom, the surge pulls divers back and forth; strong swells sweep them along even at depths of twenty metres. They tend to swing along defined "paths," being thrust forward as much as five metres, then pulled back the same distance. The effect can be quite startling, especially if the surge propels a diver to within centimetres of a solid rock face before it pulls back.

Schools of silver surfperch often "ride" the surge with divers, while individual rockfish hold their positions within the shelter of rock crevices. Foamy waves thrash overhead, tossing ragged bits of seaweed like paper confetti and tangling metre-long blades of pale green surfgrass around the rocks.

Ocean Surge

A large wave passes overhead and crashes heavily into a rocky islet. An explosion of white foam momentarily blocks my view. Five metres below the surface, the powerful ocean surge thrusts me towards the rocks. At the last minute, it sweeps me back again as the wave recedes. This is life in the surf zone, where the surge from each passing wave provides an endless roller-coaster ride. One moment I am photographing a giant green anemone, the next, I am several metres away; seconds later I am back again for another try. It is worth the effort: the top of this surge-swept reef is unbelievably rich in marine life. Colourful sea anemones compete for every protected hollow, while kelp and durable carpets of staghorn bryozoans thrive in the unprotected areas. The sloping sides of the reef are blanketed by stubby white plumose anemones and orange brooding anemones.

The giant green anemone gets its colour from a symbiotic green algae that lives within its tissues.

Above: A large lingcod rests in the shallows near a reef. An important gamefish, a lingcod can grow to 1.5 metres in length and weigh up to forty-five kilograms.

Opposite: A diver descends upon the rich shallows of a west coast reef alive with zoanthids, sponges, sea stars, and giant red sea urchins.

From Southeast Alaska to Oregon, there are vast stretches of exposed waters and outer coasts. Washington and Oregon, in particular, are noted for their pounding surf. Except for a few small islands and reefs, the coastline of both these states is completely exposed. It ranges from sharp, rocky promontories to broad sandy beaches, and from dramatic dunes to long stretches of mud flats.

In British Columbia and along the Alaska Panhandle, the western coasts of the outer islands are generally the most exposed. Uninterrupted, wild waves beat against the many granite cliffs of the Alaskan islands—mainly Chichagof, Baranof, and Prince of Wales—and the beaches, dunes, and reefs of the Queen Charlotte Islands in British Columbia. Similarly, strong breakers hammer the beaches, rocky headlands, and cliffs of Vancouver Island, carving some of North America's most spectacular tidepools in a shelf of sandstone and shale at Botanical Beach, Botany Bay.

One awesome manifestation of the frequent fury of waters off the outer coast is the lengthy graveyard of ships. Stretching from Southeast Alaska to Oregon, the havoc represents almost every maritime country and most decades over several centuries. One of the earliest logged wrecks is a two-hundred-tonne schooner, the *Lark,* which was lost in 1786 with a crew of thirty-eight near the Queen Charlotte Islands. Off Cape Flattery, Washington, lie more than 150 ocean-going ships and cargo once worth billions of dollars.

Even when the sea is calm, freak, or giant, waves have risen unpredictably to swamp ships and sweep people off shores. Although they are short-lived and relatively infrequent, these waves are particularly destructive, occurring when the movement of groups of normal waves fall into sync and mount up as huge walls of water. In 1975, a twenty-metre cresting wave struck a hydrographic launch off Vancouver Island on one of the calmest days of the summer.

Adding to the challenge of travelling these waters are the frequent thick fogs, especially during summer. The sea's prevailing northwesterly winds carry warm, moist air to the coast. When it contacts the water that is chilled by upwellings from deeper levels, the air forms thick blankets of fog that can cause chaos at sea.

Through the combined effects of fog, wind, waves, and currents, the sea has managed to reduce ships of all sizes to bits and pieces and scatter them across the sea floor. But it has also

West Coast Reef

A huge lingcod rests on the sea floor and threatens to swallow my camera if I come any closer. It is a male guarding his eggs, so I try not to antagonize him. After I take a few photographs, he suddenly blasts off across the reef after another intruder, a small kelp greenling that flees for its life. Nearby, a colourful, fish-eating anemone as large as a dinner plate lies in wait for its prey. A giant Puget Sound king crab lumbers off across the bottom like some bizarre undersea tank. Everything seems larger than real life on this wild west coast reef.

swallowed some whole, keeping them as underwater habitat for thousands of plants and animals. The steel-hulled HMCS *Thiepval*, for instance, struck a rock in Barkley Sound, Vancouver Island, in 1930 and plunged to a clam-shell seabed; soon after, it accommodated hundreds of flowerlike sea anemones and a treasure-load of other sea life.

Off these exposed shores, life is richest in the top nine to ten metres of water, a lush zone where the waves stir and serve up an oxygen-concentrated, nutrient-thick soup; below that, the abundance often drops off with dramatic suddenness. The surface life forms an exclusive club restricted to species that can survive waves powerful enough to crush them, whisk them away, or scour them with abrasive sand and pebbles.

Some plants and animals depend on the strength of their structures to resist the forces of the water. Others depend on flexibility—bending, instead of breaking, with the power of breakers. Further survival mechanisms include lying low and depending on flat shapes and smooth lines to reduce drag, or holding position by clinging tightly to a surface, clustering in large numbers or burrowing into protective places, such as holes or crevices in rocks. Some species combine several mechanisms to live in this harsh habitat, and many do more than survive—they flourish in the wild water, growing more abundantly and achieving greater dimensions on the outer coast than in other parts of the emerald sea.

Near the top of the surface zone, California blue mussels cluster closely in huge beds, their soft bodies encased in tough streamlined shells that grow as long as twenty-five centimetres. Holding fast to rocks by means of strong elastic threads, these mussels prefer life in the food-rich battering surf. So do their frequent companions, well-armoured gooseneck barnacles. They hold their places by secreting a strong substance that glues them to rocks, and their pliable stalks allow them to bend with the waves and survive the pounding.

Taking refuge in slivers of space between mussels and barnacles live many snails, limpets, crabs, worms, and other small creatures. Slow-moving sea stars feast on the tightly assembled mussels and barnacles. The sea star depends on the strong suction power in hundreds of tiny tube feet beneath its arms to maintain its hold as waves rush in.

Here in this exposed sea—and only here—grow sea palms. Resembling miniature groves of storm-blown palm trees, thickets of these plants cling tightly to rocks, their flexible, hollow stems swaying through the wildest surf. Strong waves sometimes knock the palms completely flat, but they resume their stance as soon as the waves pass by.

Deeper down other seaweeds grow, such as the reddish-coloured rough strap, named for its coarse sandpaper blades. It prefers the outer coast, as does the big bull kelp with its large floating bulb and single layer of leaflike blades. Off such exposed shores as the west coasts of Vancouver Island and the Queen Charlotte Islands grow lush forests of very tall, slender perennial kelp, a species related to the famous giant kelp in warmer water off central and southern California. Gripping with a root-like structure broad enough to seat two divers, perennial kelp favours the outer coast.

Colouring rocky offshore reefs various shades of pink, sometimes edged with white, various species of coralline algae thrive. Segments of these plants, which contain calcium similar to that found in seashells, feel as hard as the rocks they grow on, but the spaces between the segments are soft and pliable, allowing the algae to bend with the waves. Rock crust, which is a very low-growing coralline algae, literally encrusts the rocks and the shells of animals that live on them.

A large mound of feather-duster tube worms crowns an exposed reef. The flexible tubes of the worms may grow to more than half a metre in length.

36 THE EMERALD SEA

A tiny three-coloured polycera nudibranch, found only on the outer coast, crawls across hydroids and sponges.

Ear-shaped abalone, popular seafood snails that graze algae, press flat against the rocks. Their shieldlike shells protect them from rough waters off some of the most exposed shores. Lined with shiny mother-of-pearl, the tough, thick shells of northern abalone reach up to twelve centimetres in length.

Abundant colonies of colourful sea anemones populate undersea caves or rocky crevices in this wild sea. They often grow much bigger than in protected waters. Reaching diameters of twenty-five centimetres, the fish-eating anemone is large enough to use its pink or white tentacles to catch small fish. The giant green anemone—green from the populations of microscopic algae it supports—grows up to thirty centimetres in width. And white plumose anemones, looking like stately columns in a Grecian temple, reach amazing heights, standing up to ninety centimetres tall.

Another rock dweller, common in exposed waters, is the purple sea urchin, which grows to about eight centimetres in diameter. Using its sharp spines and hard mouth parts, it can bore itself a custom-sized hole in rock to escape the pounding of waves. From the shelter of this cubby, the urchin extends its many suckered tube feet to snare meals of drifting algae and debris. Some urchins further protect themselves with armour of pebbles and bits of shells, which they hold with their tube feet. Masses of sea urchins often share the same rocky area.

A great variety and abundance of rockfish also seek refuge from strong breakers on these coasts. Their characteristically large

THE EMERALD SEA

The deceptively beautiful fish-eating anemone is found only along the exposed outer coast. Fish or other small creatures that pass too close to the flowerlike anemone are captured by its stinging tentacles and delivered to its mouth.

heads and wide mouths poke out from cracks and crevices among the rocks. One species, the handsome yellow-striped china rockfish, is typical of the wild coast. Apparently curious, it often approaches and stares at divers.

Large numbers of male Steller's sea lions typically gather at rocky reefs and islands during the early summer to establish rookeries on exposed coasts. Barking and snorting, they confront each other to gain territories before the arrival of the females, most of which are pregnant. The female sea lions give birth shortly

after they reach the rookeries and mate again within about two weeks. They remain for a few more months, but the males leave by about August, after a three-month stay, to hunt food along the coast.

Unlike encounters with sea lions, which are quite common, divers seldom have the opportunity to swim near large whales. Still, the possibility exists. In the fall, great grey whales head south as far as Mexico to breed, then swim back to Alaska in the spring, completing one of the longest annual migrations of

any mammal on Earth. En route, these whales cruise fairly close to outer shores, and others, which appear to be non-breeding individuals, stay off the coast year-round. Grey whales sometimes frequent kelp beds in shallow bays along their route. Divers may be fortunate to experience the rare thrill of swimming beside one of these massive creatures—a highlight of any diving career.

Female northern fur seals also migrate up and down the coast, but they usually remain quite far offshore. Before May, they head north to the Bering Sea to breed, then after October, they travel south to California with their young. Their travels are normally beyond the territory of scuba divers who nonetheless want to know that fur seals are still out there—along with such offshore giants as the fifty-two-tonne, deep-diving sperm whale and the world's largest animal, the one-hundred-tonne blue whale. They are all part of the mystique of the exposed outer coast.

Opposite: A china rockfish swims among a colourful profusion of soft corals, anemones, and sponges.

Tidal Passages: Turbulent Waters

Calm and quiet, a narrow channel of water lies just beyond the reaches of a dense, coastal rainforest. Trees crowd the shoreline, their dark mass contrasting with the brilliant blue of the winter sky, where gulls wheel and scream. There is little to hint at the drama to come. Yet, within minutes, the relentless tide transforms the rocky channel into a raging, foam-spewing torrent.

Rising as waves that would wash easily over the head of anyone on shore, the water lashes the rocks, spinning off in eddies and dropping as whirlpools into holes a few metres deep. The surface opposite these dark pools rises crazily in domes of water thrust upwards from the irregular floor of the passage. The gulls overhead go wild and swoop low over the surface, their wings blending with the whitecaps, their cries muffled by the water's roar.

The scene is typical of many tidal passages on the west coast, where currents can make diving and boating extremely treacherous. Timing is everything, coupled with a healthy respect for the power and quick-changing nature of some of the currents. Even such strong swimmers as killer whales exhibit respect for these turbulent waters, passing through very fast-flowing channels only during slack water. Divers enter hospitable sections of many tidal passages only between current flows, and they usually keep an occupied boat on hand to meet them when they surface.

During winter, the water in these passages is often so clear that divers can see to a depth of twenty metres or so. From top to bottom and from side to side, many passage walls are completely draped in sea life. Even some of the passage floors are buried by thick carpets of wide, low-growing sea anemones. There is a profusion of colour: red seaweed, yellow sponges, pink corals, orange sea slugs, purple sea urchins, and striped rockfish. To blend in here, decorator crabs dress in strong, bold shades.

Along the inner coasts of the emerald sea, tidal passages are not uncommon; they occur frequently where a narrow water-

THE EMERALD SEA 41

Right: Cascading waters roar past a small islet in Sechelt Narrows on British Columbia's Sunshine Coast. The tidal currents in this narrow passage, known to locals as Skookumchuck Rapids, can reach speeds of more than thirty kilometres an hour.

Opposite: An undersea wall of sulphur and finger sponges, pink soft corals, and anemones greets a diver in Browning Pass in Queen Charlotte Strait, off northern Vancouver Island.

way connects two sizable bodies of water. The difference between the levels of these two bodies often produces currents, as water seeks a common level. Pacific tides that extend around Vancouver Island, for instance, arrive up to two hours apart at either end of Seymour Narrows, a short tidal passage along the island's eastern midcoast. This time gap causes a difference in water level of about a metre, creating flows of up to twenty-five kilometres per hour through the passage.

Many of the currents through passages among the Gulf and San Juan islands are also produced this way. They usually flow one direction as the tide rises, then flow the other way as the tide falls. Between times, tidal passages experience short periods of slack water, when there is little or no current at all.

The strength and speed of flowing water varies according to the particular passage. Speed is usually greatest where there is least friction—away from the bottom and sides of a passage—especially if the passage has a relatively smooth bottom and contains few obstacles. Rocks, islands, and kelp beds can all reduce the speed of the flow. Two of the fastest tidal passages in the emerald sea, and in the world, are Sechelt Rapids off the Strait of Georgia and Nakwakto Rapids at the north edge of Queen Charlotte Strait. Through these passages, water charges at speeds that sometimes reach more than thirty kilometres an hour.

Navigation in some of the tidal passages on this coast is among the most treacherous anywhere. Over the years, their turbulent waters have claimed many lives and many vessels, especially in Seymour Narrows—the most dangerous navigable channel on the coast and reputedly, one of the most dangerous in the world. Besides powerful tidal currents, Seymour Narrows experiences unpredictable vertical currents and double rows of whirlpools.

The sinking of the 1,950-tonne American sidewheel steamer *Saranac* in 1875 was the first wreck recorded at Seymour Nar-

42 THE EMERALD SEA

Sechelt Rapids

We enter the water directly below the navigational light on the rock, where only hours ago we watched a two-metre overfall of water plunging recklessly through the channel, creating huge whirlpools and standing waves higher than a man. After seeing the awesome power of the Sechelt, or Skookumchuck, Rapids at a full ebb, it is hard to imagine that anything could survive here, but a multitude of life greets us as we descend. A living carpet of anemones covers nearly every square centimetre of rock. Every crack or hole that can offer any shelter at all from the punishing currents is literally jammed with life. Looking into a narrow crevice, I can see seven red Irish lords packed side by side like sardines. They look so comical I burst out laughing, which immediately floods my face mask.

rows' infamous Ripple Rock. This particular hazard lay dangerously close to the surface in the most constricted section of the narrows and claimed twenty large ships and 114 lives until its twin peaks were blasted off in 1958. The feat is popularly touted as the "biggest non-nuclear peacetime detonation on record at the time." But accidents in Seymour Narrows continued after that because the tricky waters make navigation errors possible for even the most experienced captains; the passage claimed a large luxury cruise ship, Sundancer, as recently as 1984.

Life is more abundant in the turbulent waters of tidal passages than anywhere else in the emerald sea. Masses of plants and animals, especially invertebrates, flourish on the regular, frequent—and usually rich—supplies of nutrients and oxygen delivered by the fast-flowing currents. Acting as powerful blenders, these currents mix the water in the passages from top to bottom and create fairly constant temperatures and levels of salinity throughout. The currents also tend to sweep the passages clean, keeping them clear of silt, which, in calm waters, builds up like a smothering blanket several centimetres thick and prevents life from clinging to surfaces.

Still, the sea life must be able to survive the tremendous speed and strength of tidal currents. In the swiftest passages, species that are compact and low-growing tend to predominate; they are best able to withstand the force of the water. Generally, the passages are habitat for tough, not delicate, bodies and for those that can cling tightly to a surface or squeeze into protective nooks and crannies.

Adaptations to survival here are similar to those needed for survival on exposed outer coasts, but tidal passages demand even more. Barnacles stick especially tightly, and strong but flexible creatures, such as finger sponges, which grow more than a metre tall, bend their rubbery, yellow bodies with the currents to survive the impact of the water.

The array of sea anemones in tidal passages is astounding. Dense crowds of tiny emerald and mauve aggregating anemones drape walls and ledges in shallow water, and larger green, pink, yellow, or rust dahlia anemones inhabit crevices and trenches deeper down. Variously coloured brooding anemones with their tiny young incubating on their bases sometimes live on clusters of brilliant purple tube worms. And strawberry or club-tipped anemones cover the rocks, their reddish-pink tentacles tipped with minute white clubs.

Although these strawberry anemones carpet several sites along exposed coasts, they grow profusely only in one tidal passage in British Columbia. They have become symbolic of Discovery

A decorated warbonnet finds shelter among the delicate fans of gorgonian coral.

46 THE EMERALD SEA

A salmon-gilled aeolid nudibranch crawls slowly across a bed of white hydroids near Sechelt Narrows.

Passage, where they blanket walls as far as the eye can see, massed thickly with bright yellow sponges. Blending with this brilliance is the emerald sea's most colourful fish, the striking red Irish lord, which often props itself up among the anemones and stays so still that divers sometimes mistake it for a rock.

Discovery Passage also provides rich habitat for huge populations of one of the biggest tube worms found on the Pacific coast. Clusters of feather-duster tube worms nearly a metre tall cover wall and reef. Growing in big bouquets, dozens of flexible, leathery tubes, each two to three centimetres wide, bend with the flow of the currents. When feeding, the worms extend wide purple and red feathery plumes out of the ends of their tubes—an amazing sight.

Although plumose anemones in tidal passages do not reach the gigantic heights they do on exposed coasts, they often mass

Corals in a Cold Sea

Off a small islet in Queen Charlotte Strait, we descend past a wall covered in delicate brooding anemones towards a sight that I have heard about for several years but never seen. At a depth of twenty-five metres, we suddenly enter a small undersea canyon, and I catch my breath as I take in the wondrous sight that surrounds me: an entire field of gorgonian corals in various shades of pink and white, their delicate fans swaying gently in the slight current. Moving in close to photograph several of the twenty-centimetre-high fans, I disturb a swimming scallop. It lifts off the bottom and darts away from me, resembling a swimming set of false teeth as it bites its way through the water. One of my diving companions waves me over to where a giant Pacific octopus slowly winds its way through the maze of coral. It seems curious about these strange, bubble-blowing creatures that have invaded its domain. When my film is nearly gone, I check my air-pressure gauge and prepare to leave this beautiful coral kingdom. Suddenly, from beneath the outstretched tentacles of the octopus, a magnificent decorated warbonnet emerges, a fitting final image of a truly magical place.

thickly, creating walls of white, fluffy plumes on stalks thirty centimetres high. One such outstanding display is on Baines Bay wall in Porlier Pass between Valdes and Galiano islands in the Strait of Georgia. Among the dense crowd of stately plumose anemones that cover the wall live colourful sponges, calcareous tube worms, and hundreds of swimming scallops.

Lush beds of blue mussels grow in the shallows of many passages, attracting masses of ochre sea stars that feed on them. Not surprisingly, barnacles also grow thickly here. What is surprising, however, is that divers sometimes discover mussels, sea stars, and barnacles again at depths of ten metres—even at depths of thirty metres. Because the swift currents in tidal passages create conditions similar to those along wave-battered shores, animals that are typical of shallow exposed waters also live in deep pas-

Opposite: A colourful profusion of anemones decorates a reef in only three metres of water in a current-swept stretch of Juan de Fuca Strait. The strait, which extends out to the Pacific, is a great canyon of richly fertile waters.

A large orange-peel nudibranch approaches a basket star, which has extended its arms to feed. The orange-peel nudibranch is one of the largest nudibranchs in the world, reaching lengths of thirty centimetres.

sage waters. In fact, many thrive there. Just a few years ago, divers first discovered gooseneck barnacles at depths down to thirty metres in the super-fast flows of Nakwakto Rapids, where they grow to huge proportions, producing stalks twelve centimetres long.

Divers have also discovered blue-clawed lithode crabs in the otherwise empty shells of giant barnacles in tidal passages, such as Sechelt Rapids and Arran Rapids at the mouth of Bute Inlet, north of the Strait of Georgia. Until recently, these rarely seen crabs were thought to live only in crevices and holes in rocks along the outer coast. The crabs spend most of their lives tucked away, feeding by filtering the plankton so richly served up by fast-moving water.

One of the most beautiful—and often surprising—discoveries in tidal passages is a thick bed of pink soft coral. Because most coral lives in warm waters, many divers are amazed to encounter some species in the cold sea. Thriving in swift-flowing water, these colourful creatures occasionally anchor themselves to rocks in the wave-swept shallows of the outer coast, but they are more characteristic of tidal passages, where they grow profusely.

Within the ninety-kilometre-long Queen Charlotte Strait are some exceptional soft coral sites. The strait, which is one of the richest stretches of habitat on North America's west coast, experiences currents that are generally slower than those in some passages, such as Seymour Narrows. But the cold water of the strait supports large concentrations of plankton, feeding a wide variety of animals, including colonies of soft coral.

Browning Pass, which runs between two small islands in Queen Charlotte Strait, is a breathtakingly beautiful spot—alive with invertebrates, such as pink soft coral. Extending from depths of three metres all the way down to thirty metres or more below the surface, thick clumps of coral thirty centimetres across cover the rocks. When they extend their tentacles, the colonies of coral drape passage walls in a brilliant pink plush; contracted, the colonies resemble masses of small, red berries.

Never far away are large orange-peel nudibranchs, brightly coloured sea slugs that prey on coral. The basket star, a relative of the sea star, often wraps itself around coral, sometimes scrunching up like a ball of knotted string. If divers rub the basket star gently, especially at night, it may unfurl its skinny, multi-branched arms and wriggle them about.

On rock surfaces within the passage, pink soft coral colonies compete for growing space with many other invertebrates, such as barnacles, a dozen species of sponges, sea anemones, and tiny, plantlike purple hydroids related to sea anemones. Divers often find masses of yellow encrusting sponges and large, brown finger sponges, almost half a metre tall, pressed between clumps of coral. So rich is life on these passage walls that animals often grow one on top of another.

Related to the pink soft coral are passage colonies of firmer, more structured corals: graceful sea pens and fanlike gorgonian corals that grow smaller, but more abundantly, than the giant, metre-high specimens living in fiords. Just a few years ago, divers made the exhilarating discovery of a new pink and white species of gorgonian coral, which scientists now call *Calcigorgia spicculifera*. In more than thirty metres of clear water in a current-swept passage north of Queen Charlotte Strait, divers have encountered a vast field of this twenty-centimetre-high treasure. The discovery of life never before imagined is a special bonus that comes from diving tidal passages within the emerald sea.

Overleaf: The neon blue and yellow bands of a candy stripe shrimp contrast sharply with the bright red of its crimson anemone host. The small shrimp finds shelter along the base of the anemone, and likely feeds on scraps of food captured by the anemone.

Chapter II
Life Within the Sea

"It is the contemplation of life that inspired Father Teilhard de Chardin to meditate on the three infinities: in addition to the infinitely big and to the infinitely small, Teilhard told us there also was the infinitely complex: Life."
—ocean explorer Jacques Cousteau

Just as blue is the colour of an ocean desert, green is the colour of an ocean brimming with life. The masses of green phytoplankton that grow in the cold waters along British Columbia, Washington, Oregon, and Southeast Alaska support vast num-

Right: The long blades of bull kelp flutter in a slight current. The fast-growing kelp can reach lengths of fifteen metres in a single season, but it is often swept away by fall storms.

Previous page: One of the most brilliantly coloured of all the fish in the emerald sea, the red Irish lord rivals the most exotic of tropical fish.

bers and many species of marine life directly and indirectly throughout the food chain. In fact, this sea supports more than 600 species of plants, 4,000 species of invertebrates (spineless animals), 450 species of fish, and 30 species of marine mammals. And as the exploration of this underwater wilderness continues, more and more species are being discovered.

The nutrient supply and the temperature of the emerald sea affect the very nature of the life it supports. Its richness makes feeding relatively easy, creating many generalists that feast, often leisurely, on whatever comes their way; and its cold temperature slows the maturing process, allowing many species, such as the giant Pacific octopus, time to reach imposing sizes.

From microscopic plankton to massive whales, the infinite complexity of life is well represented in this amazing sea.

A Sea of Forests

Breaking through slits in the green-gold canopy, sunshine flickers softly on the forest floor. From this solid rock base, a mass of fast-growing bull kelp stretches up to fifteen metres tall, shading a luxuriant understorey. Taken together, the magnificent growth creates a forest as hauntingly beautiful and serene as any that grows on land.

Divers enter the kelp forest beneath the entangling canopy, its long, thick blades fluttering overhead like streamers in a strong wind. Making their way through the dense stand, they push aside rubbery stipes, the thick, hollow stalks of the bull kelp. At the top of each stipe, a gas-filled float—larger than a fist—supports the blades and lifts them up towards the light. Three to four times as broad as the float, a tough, rootlike holdfast anchors the stipe to a rock on the sea floor.

The forest is alive with fish. Several kelp greenlings, spotted blue, red, or gold, curiously trail divers as they make their way through the stand, while overhead, coppery brown kelp perch drift by. To one side, a single burly cabezon bolts into the forest to take refuge, quickly fading from view as it settles its marbled body among the vegetation. And between the rocks below, copper rockfish hover motionless.

As the divers move along the forest floor, scattered light plays on their backs, dimming and brightening in no particular rhythm. The amount of light that manages to slip through the canopy

Rockweed adorns the shallows along the edge of an undersea pinnacle.

varies continually, but the constant swaying of kelp blades means that sunlight is never kept completely from reaching the forest floor.

Here in this strangely lit world, there is a lush understorey, including several types of red algae up to thirty centimetres tall. Along with the rocks, the understorey provides safe shelter for many small animals. But as a crab crosses the sea floor close to the forest edge, the quick jaws of a spiny dogfish shark snap it up. Like many predators, this small shark finds hunting in kelp forests too restrictive, but it sometimes patrols the edge of a stand.

Besides providing refuge, the forest itself is a source of nourishment. One species of kelp, *Porphyra nereocystis*, even grows directly out of the bull kelp's stipe and nowhere else. The divers watch several different animals feeding on the plants: a small orange-spotted nudibranch, or sea slug, nibbling seaweed; a well-camouflaged northern kelp crab feeding directly on the bull kelp;

Forests beneath the Sea

The sun's golden rays flicker through the undersea forest as I swim slowly along its fringes. Spotting an opening, I enter and find myself in a small glade beneath a green-gold canopy of bull kelp. Suddenly, a solid, undulating wall of silver streams into the glade, as a huge school of herring and sandlance invades the kelp forest. I watch, mesmerized, as the fish flow around me like sparkling rivers of quicksilver. In moments they have passed by. Just as they disappear into the enchanted forest, I see a seabird fly into the school and emerge with a herring in its beak.

and the most voracious eaters of the undersea forest—red sea urchins. When thousands of these spiny creatures invade a forest, they quickly reduce the mighty bull kelp to a miserable mass of holdfasts.

Few areas of the world rival the cool, nutrient-rich emerald sea in sustaining such profuse growth of large marine plants and in nurturing so many different species of plant life, including more than thirty species of kelp alone. Most of the plants are algae, which range in size from microscopic to massive. Those, like kelp, that are big enough for the unaided eye to see easily are called seaweed. The microscopic algae—the majority of marine plant life—are mainly one-celled. Green sea grasses account for just a few of all the plants in the sea.

In these cold waters, the sea grasses flourish. Unlike seaweed, they grow roots that anchor the plants in muddy or sandy bottoms, and they produce plain, greenish flowers fertilized by threadlike, water-borne pollen. When divers swim over broad stretches of sea floor beneath quiet waters, they cross expansive meadows of pale green eelgrass. In exposed water along open coasts, they find beds of bright green surfgrass.

The leaves of both grasses support many kinds of plants and animals. In fact, one red alga, *Smithora naiadum*, which produces thin blades as long as ten centimetres, lives only on eelgrass. Small brooding sea anemones and clinging jellyfish stick to sea grass leaves, and many species of fish nibble the leaves. Some fish also hide and spawn there, and the young use the leaves as a nursery.

As divers descend the sea's upper thirty metres, they swim through what is generally a layered world of seaweed. Near the surface, most of the plants are green; farther down, they are mainly brown, then red. Some of the deepest-growing plants in these waters are the encrusting coralline algae, layers of reddish pink crust spread across the rocks. Still farther down, darkness discourages growth of most plant life.

A diver explores the edge of a kelp forest. These undersea forests teem with life; they are both a refuge and a hunting ground for a variety of species. Grey whales sometimes use kelp forests to hide their young from killer whales.

Like land plants, seaweed contains chlorophyll, which combines with light from the sun to convert nutrients from the sea to organic matter in a process called photosynthesis. The chlorophyll colours many types of seaweed green. One seaweed, the paperlike sea lettuce, has two forms of chlorophyll, so it is a particularly bright green colour. Red and brown algae also contain chlorophyll, but their other pigments mask the green. Phycoerythrin, a red pigment in red seaweed, is especially good at using the dim light that reaches greater depths.

The emerald sea supports an abundance of plant life partly because upwelling currents and turbulence regularly supply nutrients to the levels where plants grow, but also because the sea provides a wide range of habitats. On exposed, wave-battered reefs, some seaweeds cling with small, fingerlike holdfasts, while others grip with holdfasts that grow up to twenty centimetres thick and forty centimetres long.

When divers enter sheltered bays and inlets, however, they come across different species of marine plants that flourish in

A common orange-spotted nudibranch passes several orange cup corals. Well over eighty species of these colourful animals inhabit the emerald sea.

quiet waters. Here, great masses of soft, yellow-brown Japanese weed, accidentally introduced with the Japanese oyster in the 1940s, ride the surface with their many small floats. Green sheets of other seaweeds, their blades only one or two cells thick, float freely on the surface of the sea. Sometimes quiet waters also support populations of a bottom-growing red seaweed with a holdfast just two to three millimetres across. This tiny holdfast grasps a single pebble or a tiny piece of broken shell, anchoring a well-branched plant that stretches up to forty centimetres high.

Throughout the sea, reefs, rocks, ledges, caves, crevices, and forests vary the intensities of sun and shade, further increasing the diversity of habitat for marine plants. Unlike some floating plants and large kelps, which thrive in bright sun, other seaweed is very tolerant of shade. Divers commonly find velvety carpets of stunning red foliage lining dim caves and other deeply shaded areas.

Although they can explore seaweed and sea grasses, divers cannot even see the billions of drifting algae that surround them, except as an emerald green hue in the water. Forming most of the phytoplankton (*phyto* meaning plant, and *plankton* meaning drifting), the algae are mainly microscopic, one-celled plants only visible when massed densely together as a cloud in the water or a film on the surface.

However, phytoplankton is critical to sea life. Floating in tides and currents along with small and mostly microscopic animals,

Night Dive

At night, with only the beams of our flashlights to illuminate them, the creatures of the reef seem even more vividly colourful than during the day. Without the lights, it suddenly feels as though we are immersed in a sea of black ink. Then, with a sweep of my hand, I find myself floating in a universe of stars. The pinpoints of light are actually tiny microorganisms called dinoflagellates, which light up in a wonderful display of bioluminescence when disturbed by movement. My diving companion swims by; he looks like a ghost diver, his swim fins leaving a comet trail of bioluminescence behind him. Orange sea pens, which feed on the dinoflagellates, are able to glow as well. Finding one, I gently stroke it, and it lights up with the strange cold light of the deep.

A water jellyfish, one of the most commonly seen jellyfish in the emerald sea, drifts in the waters of Juan de Fuca Strait.

phytoplankton is a major component of plankton. Through photosynthesis, phytoplankton produces food for many animals, from filter feeders like oysters and clams to the blue whale, the largest animal on Earth. Phytoplankton is also a key producer of oxygen, playing an even greater role than land plants in maintaining Earth's critical supply. Some researchers estimate that phytoplankton produces as much as 80 per cent of the oxygen for this planet.

When phytoplankton drifts into sunlit waters that are especially rich in nutrients, it may grow very fast, producing extremely high populations—millions of organisms in a litre of sea water. Depending on the species, phytoplankton blooms can turn the water such colours as green, red, orange, or brown, or they may leave it clear. Much of the marine life benefits from the food and oxygen provided by these sudden blooms, but some phytoplankton blooms, popularly called "red tide," produce a poison that plankton feeders, such as shellfish, concentrate in their flesh. The poison does not harm the shellfish but it can be stored there, sometimes for long periods, harming people who eat them. Butter clams, for instance, can store poisons for two years.

At night, the blooms can create a cold light extravaganza. With every splash of a wave against a rock, complex microscopic organisms called dinoflagellates (part animal, part plant) generate chemical reactions and produce bright flashes of living light, or bioluminescence. Divers and some sea life can create the same hypnotically beautiful effect by swimming or swirling through the blooms, sometimes leaving a ghostly trail in the water behind them. Even some sea pens, when stroked, light up like green lanterns from plankton on their surfaces—just one of the strange and beautiful phenomena in this forested sea.

A Sea of Spineless Creatures

Swimming about nine metres beneath the surface of the sea, the divers barely notice a nondescript brown and black rock covered with barnacles. Suddenly the "rock" rises from its ledge. It is a fifty-kilogram octopus, which glides gracefully through the water, its eight long arms trailing behind. As the octopus moves, its skin grows smooth and gradually takes on the soft, greenish-yellow colour of seaweed-filtered sunlight.

Suddenly, there is a dark shadow, and a massive sea lion swoops down from the surface. Turning ghostly white, the octopus ejects a cloud of black inky fluid and jet-propels itself a distance of six metres towards the rocks. There it stretches thin and presses its body through a narrow crevice. Losing sight of the octopus, the sea lion shows no further curiosity and swims off.

Knowing that large octopuses tend to live a fair distance apart, the divers swim farther along, looking for a den. At the base of a rocky outcrop, they spot a pile of broken crab shells—very possibly the remains of an octopus meal. Close inspection reveals an octopus inside a den not much bigger than itself.

Opposite: Pink branching hydrocoral and purple encrusting hydrocoral are washed by strong currents in a small tidal passage. Pits on the surface of the hard skeleton of hydrocoral contain feeding polyps that have tiny tentacles to capture plankton.

The eye of a giant Pacific octopus. The octopus has excellent sight; its eye is similar to the human eye in construction, except that the pupil is rectangular. These intriguing animals are some of the most intelligent creatures found in the sea; some researchers claim their intelligence is equal to that of a house cat.

Although an octopus is not easily coaxed out of its den, the divers hope that curiosity will dominate this time. One diver extends a finger and patiently wiggles it at the den's entrance. Eventually the octopus reaches out for the finger, feels its way up the hand and then the arm of the diver. Taking hold of the octopus's arm, the diver gently eases the animal out of its den.

As the second diver snaps pictures, the octopus is drawn to the camera, moving in for a close look at the lens. It seems to be attracted to its own reflection, but that does not hold its interest for long. Pulling itself with the approximately two hundred suction-powered discs under each arm, it glides smoothly across the sea floor and disappears back into its den.

Weighing seventy kilograms or more, the emerald sea's giant Pacific octopus is the largest octopus in the world. From arm to arm, it can measure seven metres. The tips of these long arms are very nimble, able to pile up rocks and shells to help build a den and feel in holes for crabs and shellfish. These tips are also able to smell potential prey.

A marine mollusc (soft-bodied animal), the octopus has only one hard part—a sharp, black beak shaped like a parrot's beak. The octopus uses it to break open shells and tear food to pieces. To defend itself, the octopus also bites with its beak, releasing a poisonous saliva that paralyzes or kills small prey. A rough tongue inside the beak helps scrape crab meat out of the shell.

The magic of its quick colour changes is explained by thousands of pigment sacs beneath the octopus's thin skin. Tiny mus-

cles open and close the sacs, exposing colours that blend best with the background or that express its mood. Even newly hatched octopuses can change their colour.

The octopus "swims" by filling the lower part of its mantle with sea water, then squeezing it out through a siphon, which can point both forwards and back. It is the force of water leaving the siphon that propels the octopus.

Of all invertebrates (animals without backbones) the octopus has the most developed brain. It is capable of learning and remembering. Considering that invertebrates account for at least 95 per cent of the animals on Earth, the octopus occupies a very prestigious niche. Most other molluscs are very simple animals, such as snails, oysters, clams, and abalone.

But the prize for the most beautiful mollusc—and invertebrate—goes to nudibranchs. These multi-shaped sea slugs are as lovely as land slugs are plain. The bright shades and bold patterns of many nudibranchs act as danger flags to warn fish and other would-be predators that nudibranch tissue produces noxious substances.

Unlike most molluscs, whose gills are tucked away inside a mantle, some nudibranchs have gill-like projections called cerata

A giant Pacific octopus jets away from a diver by expelling water from its mantle cavity. When threatened, an octopus may also eject dark clouds of black "ink," which can confuse predators and allow it to escape.

Nudibranchs

On land, slugs are anything but beautiful, but their relatives in the emerald sea come in an astounding variety of exotic shapes and colours. Ranging from the size of a fingernail to more than twenty centimetres in length, they offer underwater photographers an endless assortment of subjects for macrophotography. On a dive in Washington's San Juan Islands, armed with several cameras, I spent well over an hour in only three metres of water—contentedly photographing nudibranchs. I never moved more than ten metres from where I entered the water, yet I captured almost a dozen varieties of multicolored nudibranchs on film such as this red-gilled aeolid.

standing up along their backs. Other types of nudibranchs breathe through cerata that are coiled around their back ends. These cerata contribute to the odd shapes of nudibranchs and explain their name (*nudi* means naked, *branch* means gill).

Among algae and eelgrass live several species of nudibranchs, such as the lovely alabaster nudibranch and the tiny transparent hooded nudibranch, which hurls its fringed hood to capture small organisms in the water. Occasionally, clusters of these hooded nudibranchs all sweep the water together.

One of the largest and most fascinating nudibranchs in the emerald sea is the twenty-five-centimetre giant nudibranch. It feeds exclusively on the tube-dwelling anemone, which builds itself a protective tube of mucous and silt, sticking only its tentacles out to feed. Divers are sometimes startled by the sight of a successful nudibranch attack on the anemone. Arching upwards like a snake, the giant nudibranch suddenly thrusts its head downward, seizing the tentacles in its jaws. As the anemone retracts, it sometimes draws the nudibranch's head right inside. As savage as it

THE EMERALD SEA 67

Right: A juvenile sunflower star creeps across a rock face, propelled by hundreds of tiny tube feet. A voracious predator, the sunflower star feeds on shellfish and whatever else it may come across.

Below: A giant nudibranch crawls across the sea floor in search of its prey, the tube-dwelling anemone.

With its shell encrusted with sponge, a swimming scallop views the undersea world through dozens of tiny eyes, visible along the top and bottom edges of its mantle.

looks, the attack is rarely fatal. After the nudibranch has eaten, it easily extracts itself from the tube, and the anemone usually regenerates new tentacles.

A potential predator, the large sunflower star, can elicit another startling movement from the giant nudibranch. The touch of the sea star is enough to start the nudibranch swimming convulsively, thrashing its body back and forth until it nearly touches head to tail. It speeds up the frenzy by beating its cerata. A few minutes of swimming is usually all that is needed to make its escape, then it settles down to hunt for more tube-dwelling anemones.

As dramatic as they are, giant nudibranchs cannot swim nearly as fast as scallops can. Some of these two-shelled molluscs are stationary, but other species, such as the spiny pink scallops, swim by opening and closing their shells rapidly, forcing water out one way and propelling themselves the other. Encrusted with purple, orange, or yellow sponge and rimmed with a row of steely eyes, the swimming scallop passes divers by clapping its way through the water like a loose pair of jaws.

The yellow sponge on the scallop may be a boring sponge, which produces chemicals that eventually disintegrate the shell. Most sponges grow on rocks and other inanimate objects, but they are all porifera (pore-bearers)—the sea's most primitive multi-celled animals.

The tentacles of a tube-dwelling anemone can be quickly withdrawn into its tube whenever potential danger is detected.

Pores are basically all a sponge has, and it has millions of them. It has no distinct body parts, not even a mouth, but the sponge draws in water and microscopic life through its tiny pores. It removes the food, then expels water through larger openings.

As passive as sponges are, they contribute to the pleasure of diving by providing vivid splotches of colour: yellow, orange, green, purple, and red. The significance of colour to the sponge is not known, nor is the reason for its wide-ranging shapes and sizes; some sponges are just specks, while others grow as tall as a diver. Many of them provide food for animals such as nudibranchs and some, such as the giant cloud sponges, provide shelter for animals, including fish.

The sea also claims a wide variety of delicate animals called cnidarians, which are mainly sea anemones, corals, and jellyfish. Their characteristic soft, tubelike bodies add colour, beauty, and grace to the sea. Divers often mistake many of them for flowerlike plants.

On rocks, shells, or other hard objects, sea anemones grow in profusion, their pinks, greens, purples, and golds blanketing rock cliffs and sea floors. Most of the time, anemones remain firmly attached to these surfaces (some live in the same spot for thirty years), but they can slide very slowly, using a muscular disk at their base. Swimming anemones can release and thrash through the water to "swim" away from predators, such as leather sea stars. Some other anemones live on shells of hermit crabs, feeding on food that floats away when the crabs eat; the anemones, in turn, help the crabs blend with their surroundings.

The giant green anemone shares a mutually beneficial relationship with the microscopic green algae that live with it, providing its colour. The algae gain habitat and protection, while producing oxygen and sugars that their host can use.

Some of the most stately anemones in the emerald sea are the white, orange, and brown plumose anemones. Their slender tentacles crown bodies that reach ninety centimetres in height and twenty-three centimetres across. Near the other end of the

Overleaf: A crimson anemone is surrounded by velvety mounds of pink soft corals.

Below: Painted, or dahlia, anemones compete for space with purple encrusting hydrocoral on a current-washed reef in Juan de Fuca Strait.

Opposite: A large Puget Sound king crab lumbers off across a west coast reef. One of the giants of the emerald sea, these crabs can grow to over thirty centimetres across.

scale, the strawberry or club-tipped anemone is only two centimetres in diameter.

Sea anemones use hollow tentacles that surround their mouths in graceful, petallike fashion to sting and trap food and to ward off enemies. Yet some predators, such as certain species of nudibranchs, avoid the sting, even turning it to their advantage. Nudibranchs that feed on sea anemones internally transfer the anemones' stinging cells to their cerata to bolster their own defence.

Another invertebrate that is immune to this venom, the colourful candy stripe, or clown, shrimp, moves comfortably among the columns, mouths, and even the tentacles of anemones that commonly sting other species of shrimp. The candy stripe, which feeds on bits of food dropping from the anemone, seems to prefer crimson anemones, but divers have also found it on dahlia, fish-eating, and white-spotted anemones.

Many coral polyps, which are very similar to tiny sea anemones, also bear tentacles with stinging cells and are sometimes

Octopus

Engrossed in photographing a colourful anemone, I don't notice I have company. The first indication is a persistent tug on my left arm, followed by a multi-suckered arm that suddenly attaches itself to my camera. I turn to find myself face to face with a giant Pacific octopus, one which has developed a strange fixation on my camera. Its rear arms are still firmly anchored in the den from which it has emerged, and it uses that leverage to try to pull the camera from my grasp. A comical game of tug-of-war ensues, but eventually the octopus relents and releases its grip. Its curiosity still aroused, it then begins to investigate my diving companion, who has been observing all of the proceedings while making excited chortling sounds into his regulator. It is my turn to chuckle now as the octopus extends a couple of arms to probe my partner's face mask. He clamps one hand over his mask to keep from losing it, and extends his other hand to gently stroke the octopus.

Its strange mottled skin, covered with bumps and hollows, seems to flush with colour when touched. Octopuses have the ability to change the colour and texture of their skin to match their surroundings, and sometimes it seems as though they do it to suit their moods as well. Abruptly, the octopus jets away into open water, and my companion follows. With its arms trailing fluidly behind it for over two metres, the giant propels itself through its ocean kingdom by expelling water from its mantle cavity. Gliding silently overhead, the octopus looks like a strange, alien ship soaring through a mystical, emerald green sky.

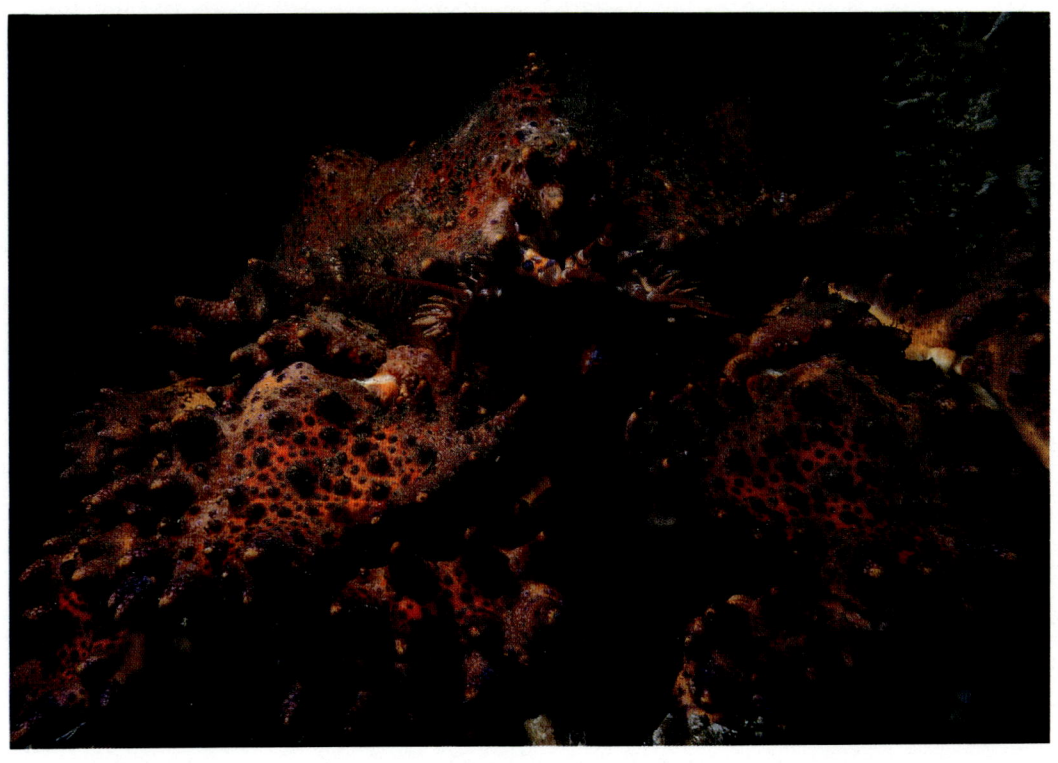

mistaken for plants. Connected by rubbery tissue, a branched colony with clusters of individual polyps of pink soft coral mass with other colonies, forming carpets of hot pink, orange, or white. The colour varies with the site. Soft coral has a skeleton of sorts (needles of hard material similar to the skeleton of a sponge), but it does not provide the rigid support found in such corals as fanlike gorgonian corals and sea pens. Although colonies of sea pens anchor themselves in sandy bottoms, they can move—however slowly—if they must.

The much less colourful, but more ethereal, jellyfish swim the sea, contracting and relaxing their wispy bells to propel themselves. On many species, long, food-moving arms follow gracefully, like tails on a kite, but jellyfish depend on their stinging tentacles to trap food.

The translucent moon jelly has about 250 fine, sticky tentacles to work with. Hanging from the rim of the bell, these tentacles trap plankton, which the arms rub off and transfer to the jellyfish's mouth. Common in quiet waters, the moon jelly is a sight, especially when viewed from above. Divers can look down on the distinctive set of four horseshoe markings near the top of its fifteen-centimetre bell.

Another sight, often startling at close range, is the giant sea blubber. With a bell reaching 60 centimetres in diameter and tentacles up to 245 centimetres long, the blubber is one of the largest jellyfish in the emerald sea. Its stinging cells can harm divers temporarily, sometimes causing skin irritations, digestive upsets, or breathing difficulties.

Unlike these soft-bodied cnidarians, crustaceans are inverte-

Above: The translucent alabaster nudibranch is commonly found in the shallows and is often seen feeding on bryozoans, minute invertebrates that form colonies of many small polyps. The alabaster nudibranch can crack open the shells of small snails with its jaws.

Opposite: An orange decorator crab hides among the columns of crimson anemones.

brates that have hard but flexible "crusts," including the many species of crabs. Characteristically, crabs have broad, flat bodies enclosed in hard outer skeletons that are either smooth, hairy, or spiny. True crabs usually have a single pair of large pincers, which they use for fighting and getting food, and four pairs of legs for walking. They mostly move along sideways by pushing with the legs on one side of their bodies and pulling with the legs on the other side.

Species in the emerald sea range from the three-centimetre black-clawed crabs often found under rocks to some very heavy-set species that are ten times larger: the brown Dungeness crab that lives among eelgrass, the tan box crab that buries itself in sandy sea floors, and the burly Puget Sound king crab that hunts among the rocks. Although the mottled colours of the mature

A profusion of painted, or dahlia, anemones covers the sea floor in Sechelt Narrows. These anemones grow most abundantly where there are strong tidal currents.

A transparent hooded nudibranch captures zooplankton in its jellyfishlike hood. Also called a lion nudibranch, it is most often seen clinging to kelp and eelgrass.

Puget Sound king crabs help them blend with their surroundings, the young crabs stand out in reddish-orange neon.

One of the oddest-looking is the heart crab, mottled green, orange, brown, and white. Named for the raised heart shape on its back, it also sports very unusual legs that are covered with long, blunt spines.

Little hermit crabs, which are not true crabs, characteristically have soft coiled abdomens that they insert into empty snail shells for protection. Species of hermit crabs in the emerald sea include the orange and the hairy, which both reach about five centimetres in length. As they grow, hermit crabs adopt increasingly larger shells, although they apparently also move house if they come across shells they prefer. Oddly, some divers have observed that mating hermit crabs often occupy the same types of shells.

Most of the spiny-skinned invertebrates, called echinoderms, are sea stars, sea urchins, sand dollars, or sea cucumbers, and of these, sea stars are the most common. The emerald sea has an abundance of them and approximately ninety different species in a wide variety of colours. Most have five arms radiating out from a central disk, but some species, like the huge metre-wide sunflower star, have more than twenty arms. More arms mean more tube feet, which protrude in rows beneath the arms and provide locomotion, and that usually means greater speed—although no sea star moves quickly. The fastest stars cover no more than about a metre a minute.

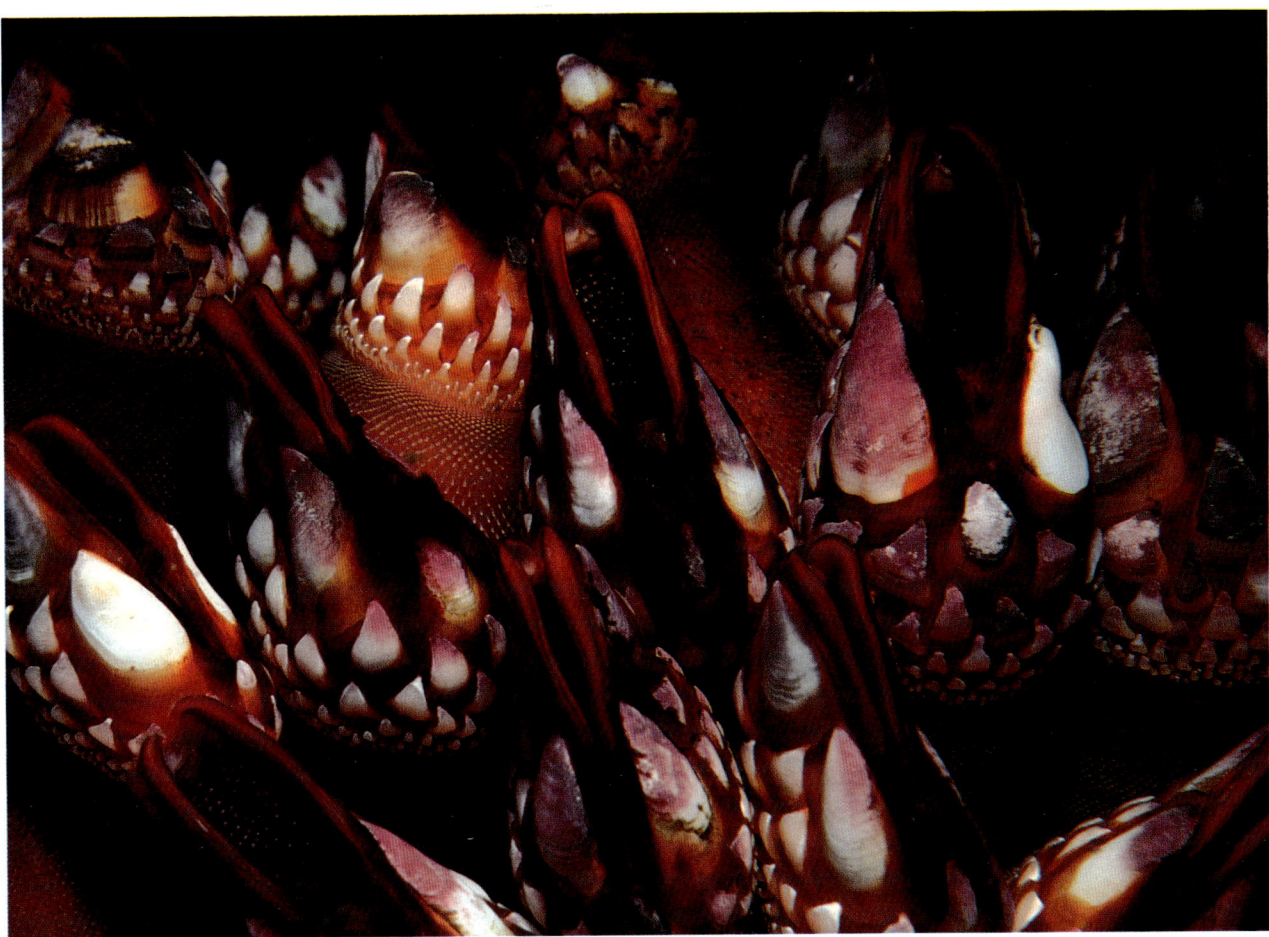

Many sea stars use the strong suction power in their feet to pull open the shells of bivalves, such as mussels and clams. When the shells gape just a bit, the star ejects its stomach, inside out, through its mouth and slips it in between the bivalve's shells. There the digestive juices begin to break down the prey's soft body. Sometimes, sea stars take in small clams or snails whole, then eject the shells and other indigestible parts.

Sea stars do not feed solely on shellfish, however. The long ray sea star, for instance, catches small fish, using little pinchers on its arms, then the tube feet move the fish along to the star's mouth for feeding. The blood star ingests mud and removes organic material from it.

Considering that some species of sea stars live at least twenty years and all can regenerate a new animal from a single arm and a bit of the disk, it is not surprising that divers see many of these invertebrates in the sea.

Most divers also encounter sea urchins, although not always happily; the sharp, movable spines on these "porcupines of the sea" can easily puncture a diver's suit and skin. Besides their many spines, sea urchins have plenty of tube feet to help move them about. These feet also catch drifting seaweed as food and help the urchins attach themselves to bottom-growing seaweed while they consume it. A special feeding apparatus, called Aris-

Above: Gooseneck barnacles, once thought to dwell strictly in the surge-swept intertidal zone of the outer coast, have recently been found inhabiting the walls of Nakwakto Rapids to a depth of thirty metres.

Opposite: A basket star unfurls its many-branched arms to feed on passing zooplankton brought in by the currents. This is the largest and one of the most unusual of the brittle stars.

THE EMERALD SEA

An orange plumose anemone adds a touch of colour to a bouquet of white anemones. Normally white, they are ocassionally found in orange or brown colorations.

totle's lantern, includes five sharp, ever-growing teeth on the urchin's underside. The most common species in the emerald sea is the green sea urchin, which is usually green but can also be white, cream, yellow, or grey.

Circular sand dollars, which grow to about eight centimetres across, are like flattened sea urchins. They, too, have Aristotle's lanterns and are covered with spines, but sand dollar spines are very short. Sometimes the sand dollars stand on edge in the sand; to help them maintain their place, young sand dollars weight themselves down, by selectively ingesting the heaviest grains of sand.

Looking roughly like the vegetable it is named after, the sea cucumber has tough, bumpy skin and five rows of tube feet that run the length of its body. Most sea cucumbers use one of two basic methods to feed. Some species, such as California sea cucumbers, move across the sea floor, mopping up mud and digesting the edible material it contains. Abundant numbers of these cucumbers can pass most of the area's surface mud through their bodies several times a year. Other species, such as the orange sea cucumber, rest among the rocks, where they take microscopic bits of food from the water.

After a successful season of feeding, sea cucumbers commonly eject their innards to rid themselves of parasites, a gesture they sometimes also make to distract potential predators. It takes the sea cucumber only a few weeks to regenerate its internal organs, making it one of the most exotic invertebrates in this strange sea of spineless creatures.

Left: The colourful tentacles of a fish-eating anemone, which can be either red or white, attract prey.

Below: A heart crab, named for the heart-shaped ridge on its back, shares a hollow in some rocks with a giant red sea urchin.

A Sea of Fish

As the motor cuts out, the sudden silence draws a wolf-eel out of her rocky den. She heads towards the anchor line where she waits, as she has on many occasions, for the splash that signals the arrival of people in her sea. Several divers plunge off the boat. Those who have been here before glance around expectantly, but not for long. Nudging one of the divers, a huge brown head suddenly appears, its wide mouth crammed with canine teeth. Its strange expression startles some of the divers, and when they see the heavy-set body that follows, they are openly frightened.

But the wolf-eel shows every sign of being friendly. She continues to press herself against the divers, circling them with her flowing body—all two metres of it. They respond by rubbing and stroking her as she moves. One diver snatches a prickly sea urchin, breaks it open, and offers it to the wolf-eel. She samples the urchin, a favourite food, then presses against the divers again, showing much more interest in being touched than being fed. All around, smaller fish swarm like flies to snap up neglected bits of urchin.

With the wolf-eel close behind, the divers head over to the rocks to relocate the den she has occupied with her mate for several years. He is bigger but, surprisingly, he is not as sociable, tending to stay in the den whenever a boat arrives. As the divers approach, the male wolf-eel pokes only his head out of the rocks, looking almost ugly enough to be cute. But some gentle hand-feeding of sea urchins coaxes him out and soon he, too, is joining the feeding.

Previous page: Strawberry anemones grow to only two centimetres in diameter, but they grow so densely that they can carpet huge sections of reef in brilliant colour.

Below: The red Irish lord is a heavy feeder, preying on animals like crabs, shrimp, and barnacles. The young may be found in tidepools.

The comical-looking grunt sculpin often hops along the bottom on its pectoral fins. Its name derives from the sound it makes when removed from the water.

The peculiar wolf-eel is just one of hundreds of different species of fish that swim in abundance in the emerald sea. Some, such as the Pacific salmon, are seen more often by nondivers than divers, but many more species are commonly seen by divers alone.

Among this diversity are fish as surprisingly colourful as many native to tropical waters. Members of the large sculpin family, for instance, spatter the sea with brilliant reds, pinks, oranges, and greens. Most of these slow-moving bottom-dwellers also sport large heads and attractive fanlike pectoral fins.

Favoured among divers for its dazzling beauty is the red Irish lord, a striking red or pink sculpin with brown, white, and black mottling. Even its big, buggy eyes contain numerous flecks of colour. Surprisingly though, its shading does not make it easy to spot. Resting among rocks encrusted with colourful algae and sponges, the red Irish lord blends well with its surroundings. Until it moves, sometimes "walking" across the bottom on its pectoral fins, divers often pass without spotting it. Even the eggs of the red Irish lord are colourful; during winter, females deposit pink, yellow, purple, or blue masses, which they guard zealously.

Although some species of sculpins, such as the green-brown and grey cabezon, reach nearly one hundred centimetres in length, several others are very small. The grunt sculpin, looking somewhat like a striped paper lantern dabbed with orange, is only eight centimetres long. Divers sometimes find one peeking out from a discarded bottle or a large barnacle shell, or they

THE EMERALD SEA

watch it swim in its peculiar hopping style. Like the red Irish lord, it also frequently "walks" on its fan-shaped fins.

Sculpins usually prefer shallow sea floors; some species even live in tidepools on rocky shores, where they adopt particular pools as their own. If waves displace them, these tidepool sculpins, coloured brown, green, or red, usually make their way back to their home pools by following their strong sense of smell.

Wolf-eel

Near Hunt Rock in Queen Charlotte Strait, the bottom drops off in a kaleidoscope of colourful sponges, anemones and corals. We are looking for friendly wolf-eels whose dens are beneath the rocks, but they don't appear to be home.

Something glides past my arm, and I am startled by a large head that abruptly peers into my mask. The head belongs to "Hunter," as the wolf-eel is affectionately known to divers, and is followed by almost two metres of sinuous, eel-like body. "Huntress" suddenly appears nearby, and I cradle her ugly but lovable head in my arms while my companion cuts up a sea urchin to feed our friends. Hunter takes the offering and returns to his den, where he munches down on it, spines and all. Several china rockfish move in to share the remains.

One of the oddest-looking sculpins is the tan-coloured sailfin. The first few rays of its upper fin are about half as long as the fish itself, and they stand straight up—like the mast on a sailboat—when the fish is still. Sometimes when it is hunting prey, such as shrimp, the sailfin may project its "mast" forward over its head and wave it back and forth. When it starts to swim, however, it usually lowers the fin. Resting or swimming, the sail-

A female kelp greenling swims past a rock covered with strawberry anemones and sponges. Kelp greenlings can grow to fifty-three centimetres in length.

fin always bears the same woeful expression created by the thick bands of black that cross its eyes and stream down both cheeks like tear-smeared mascara.

Just as attractive as some of these sculpins, many rockfish are brightly coloured and variously striped, spotted, marbled, or speckled. Sharp spines in their upper and anal fins repel predators and can pain a diver. As their name implies, rockfish normally live in rocky areas at varying depths in the sea. Although they are common fish, many species share a characteristic uncommon among fish in the emerald sea: they give birth to live young. Few species of rockfish reproduce before they are twenty years old.

With irregular yellow stripes and pale blue speckles on a background of black, the china rockfish is one of the most beautiful of the rockfish species. Divers sometimes spot one lurking among rocky outcroppings or in crevices or caves. This normally solitary fish may suddenly dash out, gills flaring and fins erect, to defend its territory from a neighbouring china rockfish.

Passing caves or crevices, a diver may spot another striking species, the tiger rockfish. Its scientific name means "magnificent with black belts," and magnificent it is—particularly when disturbed. Then the colour of its red, richly striped body, which is a warning signal to potential predators or intruders in its territory, grows even deeper.

Loose schools of greenish yellowtail rockfish, which divers often find near the surface of the sea, also have a strong sense of territory. Researchers have released captured yellowtail rockfish many kilometres from their home waters, but the fish have always managed to return, even after periods of several months.

Related to rockfish is a smaller, but equally colourful, family of fish, the greenlings. Several of its species make cooperative photo subjects. Painted greenlings, for instance, are the red and white candy-stripers of kelp beds. They hover comfortably close to divers, and their young frequently loll, as though posing, among the tentacles of sea anemones, unaffected by their stinging cells.

However, the greenling family also includes the spotted greenish-brown lingcod, which is popular as a gamefish, but not particularly attractive. Growing more than 150 centimetres long, the lingcod is, by far, the biggest greenling. It moves in swift bursts, its huge canine teeth nabbing unsuspecting fish and crabs, and when guarding its enormous mass of eggs, a male lingcod lunges aggressively at most anything that passes by. He will even grab the leg or arm of a diver and give it a furious shaking.

Besides the many colourful and attractive fish in the emerald sea, these waters are home to some bizarre species. In sunlit water, for instance, what appears to be a slender leaf may break off a cluster of eelgrass and drift away. But a closer look reveals the greenish-brown "leaf" for what it really is—a pencil-thin fish about as long as a thirty-centimetre ruler. Using a single, wispy,

A bold little painted greenling props itself up on a ridge covered with staghorn bryozoans. Juvenile painted greenlings sometimes seek shelter among the stinging tentacles of anemones.

little fin, the bay pipefish swims very slowly, and when it rests among the eelgrass, it is virtually imperceptible to predators.

The bay pipefish is the elongated cousin of the more familiar coiled seahorse. But the male pipefish does curve his body at breeding times, shaking his head at potential mates. The female deposits several hundred eggs in the male, who broods the eggs, carrying them all in one long pouch on his underside until they hatch.

Another pair of odd fish are the warbonnets. A series of fleshy projections called cirri form strange headgear that resembles the seaweed and small animals these fish hide among. Nonetheless, the reddish-brown patterns of both the decorated warbonnet and the mosshead warbonnet are quite attractive. Divers are most likely to find them poking their heads out from rock crevices or other hideaways.

As fish mature, many species change their shape and form. Few, however, change as radically as flounders do. As young fish, they swim upright; then a gradual, and peculiar, change takes place. Their skulls twist and their bodies tilt until they begin to swim horizontally. The eye from what is now the fish's underside migrates to its new topside, where the other eye is. Then the flounder spends the rest of its life as a flatfish, swimming by undulating its body and resting on the sandy sea floor.

Opposite: The exotic plumage of a decorated warbonnet, which can grow to forty centimetres in length, makes it one of the most spectacular fish in the emerald sea. It can be found in hollows, crevices, and the holes of large sponges.

Below: Sailfin sculpin amid a forest of plumose anemones. This distinctive-looking fish, with its tall, sail-like dorsal fin, is nocturnal, so it is most often seen on night dives.

The bizarre-looking ratfish swims by, flapping its winglike pectoral fins. Generally seen over sandy areas, it can be up to one metre in length.

Like many fish, including rockfish, the flounder is a quick colour change artist, altering its colour to blend with new surroundings. Triggered by visual stimuli, skin cells concentrate or disperse minute granules of red, orange, yellow, and black, creating various shades. A diver, taking a second look at what appeared to be a flounder, sees only "sand."

Skates are another family of flattened fish, but unlike flounders, skates start out this way. Because the skate's mouth is beneath its body, it cannot take in water like most fish do. Instead, water enters two openings near its eyes, then passes over oxygen-removing gills. Characteristically, the skate's broad pectoral fins join its head and body, forming "wings" that move gracefully through water. A long, skinny tail flows behind.

The largest skate in the emerald sea, and the world, is simply called the big skate. It commonly reaches two metres in length and ninety kilograms in weight, but that does not make it easy to see. The skate can match the brownish-grey sea floor so well that divers pass right over it.

As graceful as skates are, they are related to one of the oddest-looking fish, the ratfish. Its smooth, scaleless skin shimmers silver and bronze as it swims stiffly through the night, using its pectoral fins, not its tail, to propel itself. The large head, bulging green eyes, and ratlike teeth of this metre-long fish can startle a passing diver, but the ratfish presents no real danger to people.

Although most fish are bony, the ratfish and the skate have skeletons made mostly of cartilage, which is a feature they share with their better-known cousin, the ancient shark. One of the sea's most successful predators, the shark has an exceptional sense of smell and hundreds of sharp teeth. However, its reputation as an aggressive people-eater is greatly exaggerated; most species have never attacked human beings. Even the white shark—star of the fictitious book and film, *Jaws*—does not particularly target people. Like other sharks, it is an opportunistic feeder, grabbing what happens to come by, which is often seals and sea lions.

In the emerald sea, white sharks are rare, but these waters provide habitat for about thirty species of sharks, ranging from the small dogfish to the basking shark, the world's second largest fish. But the most amazing shark in these cold Pacific waters is the sixgill. Distributed almost worldwide, this broadnosed shark usually inhabits dark waters two hundred to fifteen hundred metres down, but for reasons nobody knows, it rises to shallow water—within ten metres of the surface—around

Overleaf: A red Irish lord hides in a cluster of anemones in a tidal passage.

Below: A C-O sole camouflages itself by changing colour to match its surroundings. Its name derives from the letters visible on its tail.

Vancouver Island. Nowhere else in the world is it known to visit the upper zone of the ocean on a regular basis.

Between June and November, divers most often spot these green-eyed sharks along steep rocky slopes near Tyler Rock, a pinnacle reef in a sheltered part of Barkley Sound, and by Flora Islet off Hornby Island in the Strait of Georgia. At times during July and August, divers may see the Flora Islet sixgills as frequently as once on every second dive.

Appearing as a hulking shadow, usually between 1.5 and 3.5 metres long, the sixgill shark "rides" the sea cliff up from the bottom where it likely feeds on crabs and fish. Generally, it swims slowly, almost dreamily, but sometimes it can take off in sudden bursts of speed.

Named for the number of gill slits it has (most shark species have five), the sixgill is one of the most primitive and puzzling sharks on Earth. One possible explanation for its surprising movement to shallow water may be that females may come to give birth where food is plentiful and predators are few (the pups are about seventy centimetres long at birth), but that does not explain the presence of adolescent sixgills in the shallows. The many mysteries surrounding the sixgill shark remain.

The emerald sea's biggest fish is the enormous basking shark, reaching fourteen metres in length and weighing more than three tonnes. But as big as it is, the basking shark is not a fearsome predator; oddly enough, it feeds on plankton. Its teeth are predict-

Sixgill Shark

Twenty-five metres below the surface of Barkley Sound, a great hulk appears out of the gloom and swims slowly towards two divers, who hover near the base of an undersea pinnacle. Over three metres in length, the shark moves through the water with a powerful grace that sends shivers of excitement rippling through the divers. It is a sixgill, an ancient shark whose ancestors have cruised the world's oceans for millions of years. But this encounter is unusual, for it is only here, in a few remote sites along the coast of British Columbia, that this deepwater shark is known to inhabit the shallows for a few months each year. The reasons for this phenomenon are still unexplained, but it was the possibility of an encounter with one of these huge, sluggish sharks that lured the divers here. Now, as the shark approaches them, they apprehensively push off from the rocks and begin swimming alongside the slow-moving beast.

The bottom-dwelling cabezon, the largest member of the sculpin family, can grow to nearly a metre in length.

THE EMERALD SEA

ably small, but the mouth and gill slits of this shark are large to accommodate the intake of great volumes of water. The gills are equipped with elaborate, long gill rakers—bristlelike sieves—to strain the food out of water. Divers sometimes spot a sluggish basking shark feeding near the surface.

One of the sea's most common sharks, the little spiny dogfish, often swims alongside divers, but it does not attack people. Known to live up to forty years, dogfish spend much of their life on the move, sometimes in large schools and often making amazing migrations. One dogfish tagged off Washington was caught north of Japan eight years later—a remarkable feat, but just one of many that take place in this cold sea.

A Sea of Mammals

A small boat rides the offshore waves alone, when suddenly, several Pacific white-sided dolphins appear on either side of it, riding in unison with the bow. Before long, the rest of their immense herd shows up: three hundred to four hundred glistening black backs spread in every direction as far as the eye can see.

At least a dozen of the dolphins swim right with the boat while some play in its wake. Others leap clear of the waves and flip onto their backs, completing as many as fifteen somersaults in quick succession.

When the boat stops, so do the dolphins. Quickly, the divers gear up and fall back into the water. As they hit, a dozen sleek

Above: A male scalyhead sculpin, living in an abandoned barnacle casing, puts on an elaborate display to warn off another male and perhaps attract a female.

Opposite: The striking china rockfish is an inhabitant of the outer coast. It is highly territorial and usually stays close to a place to hide.

bodies, each weighing more than 135 kilograms, move up for a close look. The divers swim slowly, watching the dolphins looking back at them. Some of the animals lose interest and start to move on, but as soon as the divers begin to "play," the dolphins are back.

Slapping their plastic fins together, the divers swim, spiral and spin, awkwardly imitating the dolphins. The animals respond, moving three or four abreast in formation with the divers. As the excitement rises, so does the dolphin's chattering, chirping, and clicking.

Even when the divers surface, the dolphins surround them, taking the opportunity to expel stale air through their blowholes. Then, their small dark eyes glancing sideways at the aliens in their sea, the dolphins grab a fresh breath and submerge. Although the shore is a distant refuge, the divers feel no fear moving among these speedy mammals. As lively and as active as dolphins are, they display no aggression towards people.

Pacific white-sided dolphins have a reputation for being acrobatic and sociable. They commonly move in herds of about one hundred—occasionally up to twenty-five hundred—and sometimes they even swim with other marine mammals, such as sea lions. Since divers began to swim with members of large herds off northern Vancouver Island in the last few years, the dolphins have appeared to welcome the company.

Grey Whale

Slipping quietly over the side of the zodiac with my camera, I swim out to try to place myself directly in the path of the advancing grey whale. The whale blows directly in front of me, and I take a deep breath and dive to meet it. The water is only about three metres deep, but I can feel my heart pounding as I drift just off the bottom, waiting for the leviathan to appear. It suddenly materializes out of the emerald green haze and a few seconds later I am swimming alongside a ten-metre-long grey whale. Its huge eye regards me curiously, but it does not slow or change direction. With a final movement of its massive, barnacle-encrusted tail, it disappears into the gloom.

Although they are abundant in the emerald sea, Pacific white-sided dolphins migrate annually, often wintering in warmer waters to the south. Scientists have found tiny grains of magnetite in their ears, which may mean that dolphins use electromagnetic radiation from Earth's magnetic field to guide their journeys. The discovery may help explain why some whales beach themselves. Using geomagnetic maps, scientists are finding significant magnetic irregularities at places where beaching often occurs.

The Pacific white-sided dolphin is just one of many kinds of toothed whales that live in this sea, including other species of dolphins, porpoises, and what is believed to be the largest concentration of killer whales in the world.

One or two Dall's porpoises, the fastest of the small whales, also ride the bow waves of boats at times, but they tend to tire of it more quickly than Pacific white-sided dolphins do. The black-and-white Dall's porpoise is more often seen zooming by before disappearing in a blur. By contrast, the little 1.5-metre-long harbour porpoise never approaches boats and almost always surfaces inconspicuously. Although it lives close to shore, divers seldom see the harbour porpoise.

Starting in 1987, there have been occasional sightings of false killer whales, which are primarily warm-water mammals, along the outer coast and in protected waters such as Puget Sound. One of the whales seemed to be quite sociable, following a small ship in Barkley Sound each day and occasionally approaching divers. The same false killer whale was later spotted in English Bay off Vancouver, where it appeared to live alone, periodically thrilling people on shore with its acrobatic leaping.

Still, the most popular whale is likely the killer whale, or orca, which is a much more common sight along the coast. Several of these black and white beauties travelling through the water together make an impressive show. They usually move and hunt

A pod of killer whales cavorts near the surface of the emerald sea. The resident population of killer whales along this coast is among the best-known and most-studied whale populations in the world.

in stable family herds called pods, led by their mothers.

Resident pods of five to fifty killer whales roam about eight hundred kilometres of the coastline, feeding on fish, while transient pods of up to seven killer whales cover about twice that distance and feed mainly on seals and sea lions. Both resident and transient killer whales use pod-specific dialects to communicate within their families, but resident pods, which are more sociable, seem to "talk" more often and use a wider range of sounds than transient pods do.

In 1990 and 1991, sightings of new pods of killer whales in British Columbia surprised and excited whale researchers. Spotted mostly off the Queen Charlotte Islands, the new pods resemble resident pods of the south coast. Loosely called the "offshore whales," these killer whales appear to be living in pods of twenty-five or more and feeding on fish, but they remain a mystery.

When divers are underwater, they sometimes hear killer whales communicating with each other, but the pods do not usually approach. Occasionally, however, these whales display some curiosity about divers and their scuba gear, leaping playfully over bubbles that rise to the surface.

Whales without teeth, such as the grey, minke, and humpback whales, form the other broad grouping of whales. They filter plankton and small fish from the water using baleen—several hundred long plates, which hang from the upper jaw and feel somewhat like fingernails. The bristly inside edge of each plate helps snag food when the whale forces water back out through the baleen. Then the tongue scrapes off the food and hurls it back to the throat.

Some of these baleen whales consume tonnes of food a day. Because they feed directly on plankton and small invertebrates, they are able to acquire large amounts of food with little energy and grow to the sizes they do. Whales often skim plankton along the surface of the water, but the great grey whale strains out mouthfuls of the sand and mud that it scoops off the bottom. In shallow water, its big thirty-tonne body sometimes makes trails in the sand.

Most grey whales spend about half their lives migrating every year between Alaska and Mexico. Travelling alone or in groups of up to six-

A sea otter uses a rock to break open a sea urchin, one of its main sources of food. Once virtually wiped out throughout much of its range, the sea otter has been reintroduced to parts of the coast and is making a comeback.

teen, they swim steadily, usually staying within a kilometre of shore. Some of them seem to enjoy rubbing their barnacle-covered bodies against small boats. For a diver to join them in the water is a rare but intensely exciting experience.

Occasionally, divers see a nine-tonne minke whale, the smallest baleen whale in the emerald sea and one that is much harder to find than the grey whale. Sometimes the minke swims along with a boat, but it spends very little time at the surface. Usually travelling alone, it occasionally feeds with killer whales.

Although humpback whales normally live in offshore waters, they often appear in the protected, inland waters of the Alaska Panhandle. Sometimes they hunt shrimplike krill by encircling them with nets of bubbles. Then the whales charge up through the "nets," mouths held wide open, and gulp down the krill. Male humpbacks are also known for their amazing songs, which may be linked to mating or to warding off other males. Singing more loudly than any animal on Earth, they form patterns of sound that last thirty minutes each and are repeated for hours.

Along a few protected stretches of the coast lives a much smaller marine mammal, the sea otter, which is cousin to such land mammals as river otters, weasels, and minks. Although sea otters lack the strong diving and swimming skills of other marine mammals, they are the only ones able to use a tool.

Diving to the sea floor, the sea otter uses its dextrous front paws to grab a small rock and such food as clams. At the surface, it rolls onto its back, places the rock on its chest and cracks the clams by tapping them on the rock. About every thirty seconds, the sea otter clutches the rock and uneaten food and rolls over, washing away the debris.

This constant flushing of debris is not just a nicety. The sea otter depends on its fine, very dense fur to keep warm in the cold sea. If the fur is dirtied, the coat quickly loses its insulating and waterproofing features, and the otter dies. The animal grooms continually, "combing" its fur with its forepaws, blowing air into its coat and turning somersaults to wash away debris and puff up the fur.

Sea otters eat, relax, and sleep on their backs so they can hold their limbs—the less furry parts of their body—up out of the cold water. A female sea otter will also cradle her golden pup on her

A shy harbour seal peeks out from behind an old dock piling. Harbour seals are often observed near fishing docks, where they sometimes feed on discarded fish parts and on the fish that congregate beneath docks.

Pacific White-sided Dolphins

Since I was a boy, I have always dreamed of swimming with dolphins, and now I am floating ten metres beneath the surface of Queen Charlotte Strait, watching excitedly as several dozen Pacific white-sided dolphins approach us from the depths. The beautifully streamlined creatures surround us, as fascinated with us as we are with them. Swimming in unison, they arc and spin in a beautifully orchestrated dance. Kicking my feet together, I try to dive and spin through the water as they do. I make a slow and clumsy dolphin, but suddenly I am joined by three dolphins who arc and swim in unison with me, no more than an arm's length away. Looking into their smiling, intelligent eyes, I wonder if they are trying to teach me to swim like them. Out of breath from the excitement and the exertion, I finally have to stop. One of the dolphins turns and comes back to me as if trying to coax me to continue, but finally it swims off to join its companions.

chest. She grooms the pup as intensively as she grooms herself, leaving it only long enough to get food. Then she floats the pup on its back, often wrapping some kelp around it to keep it from drifting off.

To satisfy their huge appetites, sea otters spend a good part of their lives hunting; they consume a quarter of their body weight every day. Over a lifetime, some eat so many red sea urchins—one of their favourite foods—that their teeth and bones turn a purplish-pink. But this dedicated foraging is an inadvertent blessing to kelp forests. Where sea otters keep the numbers of voracious kelp-eating urchins in check, the forests flourish, boosting invertebrate and fish populations, and in turn, large numbers of birds and mammals.

Besides whales and sea otters, a third group of warm-blooded marine mammals swims this cold sea: sea lions, seals, and fur seals, collectively called pinnipeds, or fin-footed animals. Well-adapted to cold temperatures, their bulky bodies are insulated with deep layers of fat wrapped in thick skin, and their circulation is lower wherever body heat would be lost to cold water.

Like whales, pinnipeds produce a variety of sounds to communicate and to find each other. Some species, such as the California sea lion, also use sounds and echoes to help locate prey when they are deep-water fishing.

All pinnipeds are excellent swimmers and divers. As soon as their smooth bodies hit the sea, their heart rates drop, their blood vessels constrict, and their lungs begin to collapse. Blood flows as usual to the brain and heart, but less reaches non-vital organs, such as kidneys. To stay underwater, pinnipeds are able to draw on oxygen in their muscles as well as in their blood.

Steller's sea lions are the clowns of the sea, often diving and pirouetting around slow-moving divers. Their large eyes help them see in plankton-filled waters.

The water skills of pinnipeds are a distinct advantage when they are hunting food or escaping enemies, but their constant swimming and diving also appear to give them pleasure. Sea lions frequently explode into action, chasing each other, tracking bubbles, and tossing seaweed about. One sea lion will occasionally grab a piece of kelp and flick it at another, only to have it snatched away by a third sea lion joining the game. Although the antics of sea lions sometimes resemble a performance of circus clowns, they are amazingly graceful swimmers and divers, especially considering their bulk (a male Steller's sea lion weighs about a tonne).

Sea lions frequently extend their playful nature to divers. These mammals like to swim with the people who appear in their sea, getting especially excited if the divers strive to keep up. Whizzing about at high speeds, sea lions sometimes will race straight towards a diver, then veer off at the last minute. With their big flippers extended as wings, they arc and soar like fighter planes, their raucous barking shattering the silence that usually surrounds divers underwater.

Often swimming in pairs or groups of a dozen or more, sea lions may try to goad divers into playing, settling near them on the sea floor and heckling them by chewing at their plastic fins or by hanging upside down and blowing bubbles at their face masks. Then, if nothing much seems to be happening, the sea lions go about their business.

Although harbour seals, the most common species of seal in the emerald sea, also display an interest in people, they tend to be more reserved than sea lions. Seals are more likely to trail divers curiously than to play with them. In bays and inlets all along the coast, harbour seals frequently haul out onto shore and bask in warm sunshine, but like all pinnipeds, they spend much of their time swimming and diving. Harbour seal pups can even swim at birth.

The water champion among pinnipeds in this sea is the elephant seal, but it often swims far offshore and dives into very deep water. Researchers recently recorded an elephant seal reaching depths greater than twelve hundred metres—about forty times the depth that a scuba diver normally reaches. Weighing about two tonnes each, male elephant seals also hold the pinniped record for size. These giants appear quite often in Georgia Strait-Puget Sound and Barkley Sound.

Another magnificent diver, the northern fur seal, named for its dark, very dense fur, seldom encounters divers. On its long annual migrations between Alaska and California, it rarely comes within twenty kilometres of the coast—remaining a casual, but striking, visitor to the emerald sea.

Overleaf: The wheelhouse of a small shipwreck is garlanded with plumose anemones.

Chapter III
People and the Sea

"A man is rich in proportion to the things he can afford to let alone."
—*naturalist, essayist, and poet*
Henry David Thoreau

In a relationship that is centuries old, people have looked to the sea as a highway for travel, a reservoir of food, and an abiding source of mystery. The Pacific's cold emerald sea is no exception. Vessels of all kinds sail its surface every day, while beneath the waves, shipwrecks hold stories and secrets of the past. Vast communities of plant and animal life thrive in the rich

Previous page: Their epic journey almost complete, these sockeye salmon have made their way from the ocean to this small stream over a thousand kilometres inland. Here they will spawn and then die.

Below: Disturbed by a fish that swam too close, this branch of pink soft coral has retracted the numerous tiny tentacles of its polyps, which it normally extends to feed on plankton.

water and varied habitats of this sea, which also supports species yet to be discovered.

The more people explore this underwater wilderness, the more they realize there is much to learn about the sea—and to learn from it. The challenge now is to preserve the quality of its waters and the abundance of its life. Compared to the ocean off the eastern coast of North America, the emerald sea is relatively pristine. Respect for the needs and nature of this sea will help preserve it, but that calls for a change in attitude. More than ever, we must learn to be rich in Thoreau's terms: rich in proportion to the things we can afford to let alone.

A Sea of Riches

One dull fall day, several research divers entered the water at the approach to Becher Bay in southern Vancouver Island. Months of fruitless surveys had left them solemn and discouraged, but a descent of just twelve metres in these clear waters

A diver peers from one of the open hatches on the *Chaudiere*, sunk off Kunechin Point, nine kilometres north of Sechelt, British Columbia.

drastically changed their mood. Beneath them sat exactly what they had been searching for: masses of large, round stones that represented a traditional aboriginal fishing technique hundreds of years old.

Using nets, which they anchored with ten to twelve stones at each corner, the Salish people had created artificial reefs close to shore. When migrating salmon swam across these "reefs," men in canoes pulled the nets upwards and trapped the fish. Every season, they laid new anchor stones, using the same sites again and again. Researchers estimated that, between the fifteenth and eighteenth centuries, a total of about 62,000 stones had been placed in Becher Bay. Believed to be the first comprehensive analysis of an underwater prehistoric site in Canada, this 1984 study revealed how early west coast people reaped some of the rich fish resources of this sea.

Another study demonstrated the effectiveness of early harvesting methods of a very different resource: wampum shells from tiny molluscs called dentalia. For twenty-five hundred years, aboriginal people in western North America used the plain, tusk-like shells as currency. The longer the shell, which grows up to five centimetres, the greater its value.

Theoretically, harvesters worked from a canoe, using a broom-like contraption made of wood splints weighted with rocks and stuck on a pole twenty-one metres long. They jabbed the tool into the sandy sea floor where the dentalia burrowed, capturing the little shellfish between the splints.

Above: The ringed top snail, with its alternating bands of purple and gold-yellow, is one of the most striking of the shellfish found in the emerald sea. About three centimetres tall, it feeds principally on hydroids.

Opposite: A striped seaperch swims past an undersea wall of stately plumose anemones. Striped seaperch school in shallow waters and graze small animals from rock outcrops.

In 1991, researchers headed for a rich dentalia bed in Kyuquot Sound on Vancouver Island's west coast to test the technique, doubted in some scholarly circles. Eighteen metres below the surface, divers watched the tips of the dentalia poking out of the sand as researchers above used a tool built according to ancient design. As it sank ten centimetres into the sea floor, several dentalia became wedged between the wood splints, proving the tool's effectiveness even in deep water.

Although early people tended to use the sea's riches simply and conservatively, visitors and later inhabitants established a much less enviable record. Over many decades, they ruthlessly slaughtered sea otters and fur seals for their fine fur and whales and elephant seals for their blubber. Europeans also exploited toothless whales for their strong, supple baleen—material used in various products, such as buggy whips and skirt hoops. In more recent times, military pilots used killer whales as targets for bombing practice. Even today, rifles are sometimes turned on killer whales, seals, and sea lions to guard commercial fishing operations, and harbour seals, often caught in fishing nets, may be clubbed, then tossed overboard.

Recognition that the survival of marine mammals was threatened gradually brought about legislation and international agreements to protect them, and populations started to increase. The numbers of fur seals rose throughout their range, and during the 1960s and 1970s, biologists transplanted sea otters from parts of Alaska and California to Southeast Alaska, British Columbia,

A diver explores Browning Pass in Queen Charlotte Strait, off northern Vancouver Island. Sulphur and finger sponges, pink soft coral, and anemones abound in these waters, where strong currents ensure a rich supply of nutrients.

Washington, and Oregon (though Oregon appears to have no survivors). Populations of whales increased slowly. In 1992, the United States removed the grey whale from its list of endangered animals, but many whale species, such as the humpbacks, have achieved just a fraction of their former numbers.

Although there are plenty of fish in the sea, the means of harvesting this rich resource has changed drastically from the simple techniques of centuries past. Fishing has become a science and a massive, high-tech industry that employs radar, sonar, computers, and underwater cameras to target its quarry. Unfortunately, the zeal to harvest this "limitless" resource has resulted in serious depletions of populations of fish, such as herring, especially during the 1960s and 1970s.

Efforts to protect economically valuable fish, such as herring and salmon, have sometimes been devastating to other species. For instance, the big, but docile basking sharks, which occasionally tangle themselves in fishing nets, came under vicious attack in the 1950s and 1960s. With a specially designed, razor-sharp knife blade fitted onto the bow of a fisheries boat, crews managed to slice these floating plankton feeders in half. Between 1955 and 1969, 414 basking sharks in Barkley Sound died this way. Other patrol boats deliberately rammed the sharks, killing another 200 to 300. Today, the basking shark is rarely seen in the sound.

Sharks generally experience a slow rate of maturation and reproduction, making their populations highly susceptible to disaster from hunting or fishing. Because of this, scuba divers and scientists expressed concern about an experimental commercial shark fishery set up in 1992 in British Columbia, especially as it targeted the sixgill shark, among others. This deepwater species, which appears near the surface in British Columbia, gives divers and researchers the rare opportunity of seeing a spectacular fish from another ocean realm. Given the lack of understanding of the basic biology of the sixgill and most other sharks, those protesting the fishery were deeply concerned.

Today, public support seems to be relatively easy to rally when the object is to save marine mammals or fish. However, there is seldom much sympathy for the spineless riches of the sea, the invertebrates. The few that do receive public notice are usually

Overleaf: A tiger rockfish is intrigued by a diver's light. These solitary fish are often found in caves and crevices.

Below: Creeping pedal sea cucumbers spread their bright red tentacles to feed on zooplankton. Crimson anemones and zoanthids cling to the rock above.

A male wolf-eel shares its den with a tiger rockfish. The wolf-eel is not a true eel, but is actually an elongated member of the wolf fish family.

those that are popular as seafood and that are becoming increasingly valuable to the economy of coastal communities. Some, such as oysters, are raised through licenced aquaculture operations, but those that are harvested from the wild frequently suffer harmful consequences. Poaching, for instance, is common among invertebrates, such as octopuses. In 1990, researchers had to cancel their study of octopuses in Puget Sound when the population at the site dropped from more than seventy-five to four in the project's third year. Poaching for the restaurant market is believed to be the cause.

Scientists gathering in Vancouver, British Columbia, in 1993

Submarines

I have often dreamed of descending deep into the abyss in a submersible and being able to linger for hours, photographing the deep-water gorgonians and other wonders that dwell beyond the safe reach of scuba diving. Perhaps someday I will be able to, for the technology exists, much of it developed right here on this coast. Vancouver is known as one of the largest producers of civilian submarines in the world. A wide range of submersibles has been developed and built here, from the Atlantis class of tourist submarines now in use in tropical destinations around the world, to the Pisces class of deep-diving research submarines. The high-tech newtsuit is a one-atmosphere deep-diving suit, which allows divers to safely work at depths of over three hundred metres, encased in a hard shell of protective armour. Remotely operated vehicles known as ROVs can operate at great depths, relaying high-resolution video images back to the operators on the surface. With tools such as these, we can now explore the deep reaches of the emerald sea and find many as yet undiscovered wonders.

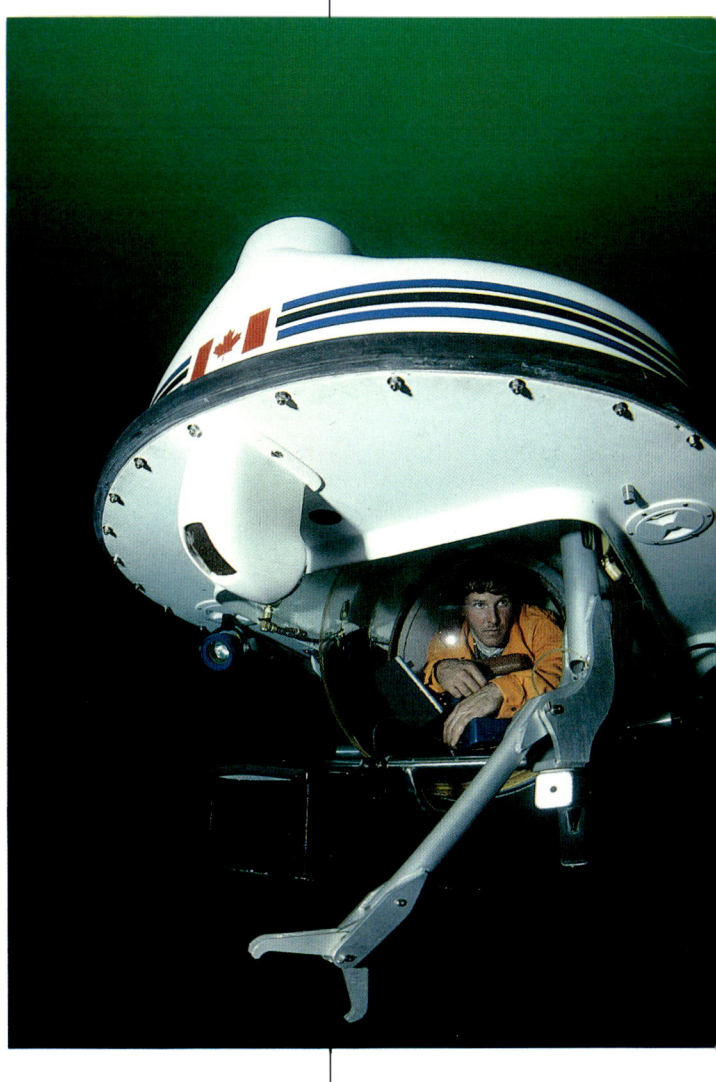

for a meeting of the American Society of Zoologists expressed strong concern that the public is generally ignoring and destroying invertebrates of all kinds. Both on land and in the sea, these animals are being wiped out at an alarming rate—sometimes disappearing before anyone documents how they fit into the web of life. Added to the decline in fish and marine mammal populations through overharvesting and habitat destruction, the loss of invertebrate life threatens to make people victims of their ignorance as well as their greed.

A Sea Worth Saving

Looking carefully through a bed of mussels, a scientist extracts an attractive shell belonging to a whelk, a common snail that feeds on shellfish and barnacles. Close inspection of this five-centimetre-long snail reveals a gross distortion: a large, deformed

Opposite: A diver and a male kelp greenling drift through a small undersea canyon.

penis growing out of the back of its head. What is equally strange is that the whelk is female. Unlike other individual snails, which have both male and female sex organs, whelks are normally one sex or the other. Clearly, the whelk in the scientist's hand is not nature's design.

The scientist is one of several studying whelks on the coasts of Vancouver Island, the mainland coast along the Strait of Georgia-Puget Sound and the outer coast of Washington's Olympic Peninsula. As females became afflicted with this strange phenomenon, whelk populations near harbours and marinas had been declining. Only at one location within the study area were whelks unaffected and that place had no boat moorage. The bizarre distortion of whelks, which are extremely sensitive to habitat pollution, appeared to occur where they came in contact with tin-based paints on boats.

By the late 1980s, regulations had come into effect restricting the use of this boat paint. Now scientists want to see if the effects on Pacific coast whelks are reversing as they have in France, where regulations controlling the paint were brought in about five years sooner.

The effects of pollution on sea life can be more catastrophic than the effects of harvesting. Populations can be reduced faster, and if habitat is seriously damaged or lost, species can suffer enormous losses, possibly disappearing altogether.

Ocean pollution takes several forms. During the past century especially, the sea has become a receptacle for wastes from industries, communities, and ships. While legislative efforts to reduce this pollution have been and are being made on both sides of the border, the quality of coastal waters in some areas has been hard hit. Increases in the numbers of beach closures and decreases in the sizes of the shellfish harvests are two of the most visible and frequent results. Filtering forty-five litres of sea water every day, the mussel shows one of the earliest signs of trouble. Scientists measure pollution levels in the tissue of this widely distributed shellfish as part of a worldwide pollution alert program called Mussel Watch.

Poor health of fish is another sign of low water quality. Governments have been introducing regulations concerning the discharge of chlorine compounds from pulp mills to reduce the deleterious effects on fish. However, some researchers have found that fish have been getting sick when exposed to effluent from mills that produce unbleached pulp and do not discharge chlorine compounds. They suspect that a chemical found in all pulp mill effluent may be the culprit. Public concern over this research is amplified by the discovery that chemicals from pulp mills seem to be widely dispersed in the Strait of Georgia, even in the deepest portions.

The effects of more industrial activity and a rapidly growing population on water quality in the Strait of Georgia-Puget Sound are serious problems. The region is expected to double its present population of 5 million people within twenty-five years. During

Sinking of the *Chaudiere*

Looking down from the cliff top, it seems as if the entire bay is littered with boats. Hundreds of people have shown up on this crisp winter morning. In the centre of the bay, listing now to its port side, but still towering over all the pleasure boats, floats the 112-metre-long HMCS **Chaudiere***. Formerly a Canadian naval destroyer escort, the once-proud vessel is about to take on a new life as Canada's largest artificial reef. Horns and cheers echo across the bay as water begins to flood into the holds. The ship lurches even further to the port side and the stern begins to sink. The massive bow guns suddenly swing and point downwards, and in an explosion of spray and bubbles the* **Chaudiere** *disappears beneath the sea.*

Later, approaching the ship underwater, I am awed by its massive size. It dwarfs a pair of divers exploring the smokestack, and as I peer into one of its dark holds I feel as though I am staring into a huge cave. We swim towards the bow and come across the large bow guns, pointing silently now into the depths. A lone rockfish hangs nearby, the only sign of life we have seen so far. Given time, however, the wreck will become a haven for marine life. In a few years anemones and other creatures will garland its hull, and schools of fish will seek food and shelter in its numerous holds.

1993 workshops on its future, representatives from governments, businesses, labour, and regional interest groups expressed concern about several ocean-quality issues, including sewage discharge.

Washington cities, such as Anacortes, Bellingham, and Port Angeles, which all empty sewage into the sea, have installed secondary treatment facilities. However, Victoria, on Vancouver Island, drains minimally treated sewage into the strong currents of the Strait of Juan de Fuca and has rejected proposals for further immediate treatment. The decision created a furore, but with scientific reports in hand, Victoria administrators claimed that environmental effects from the discharge were minimal, affecting only a small area near the outfalls. The municipality monitors the effects of its discharge and plans to implement further treatment in stages. It is also developing a program to reduce the discharge of toxic contaminants into sewers.

One of the most pervasive and devastating forms of pollution is the oil spill. Unfortunately, three sizable spills have threatened the emerald sea in very recent times. In 1988, the fuel barge *Nestucca* collided with its own tug near Grays Harbor, Washington, and spilled about eight hundred tonnes of oil. Winds and currents carried the syrupy liquid north past Cape Flattery and Vancouver Island, killing more than thirteen thousand seabirds and smothering sea urchins, limpets, snails,

barnacles, mussels, clams, and oysters in its path. Even five months later, harvesters were removing oiled crabs from their traps.

The following year, the supertanker *Exxon Valdez* caused the largest oil spill ever off the west coast of North America. The ship ran aground and dumped up to forty thousand tonnes of oil into Prince William Sound. The oil flowed along remote shorelines to Southwestern Alaska, killing an estimated 600,000 seabirds and about 5,500 sea otters. Several killer whales in the area were missing from their pods shortly after the spill and are presumed dead.

The *Exxon Valdez* spill, which occurred at the outer fringe of the emerald sea, is a grim example of what could happen if a supertanker were to dump oil farther south. In fact, in 1991, when a freighter struck the Japanese ship *Tenyo Maru* in the Strait of Juan de Fuca and created a minor oil spill, residents of the emerald sea's most populated coast were stunned. That the horrors of an oil spill could occur practically in their backyards was clearly driven home.

Most oil spills are small, often too small to notice, and not all represent leaks. Oil also enters the sea as waste from municipal storm drains and as bilge from boats, but all of it wreaks havoc with most animals that ingest or inhale the oil or become coated with it. The ills escalate as life throughout the food chain is negatively affected and as habitats, such as eelgrass meadows and muddy sea floors, are ruined.

Each week, three to five oil tankers pass through the Strait of Juan de Fuca to refineries, such as Cherry Point near Anacortes, Washington, and tankers will likely sail these waters for years. One study by British Columbia and the states along the emerald sea predicts that a major oil spill will occur every twenty years along this coast.

Industry and government have been developing ways to cope with the next spill. However, the effectiveness of cleanup methods depends upon such highly variable conditions as weather, ocean currents, the nature of the shoreline, and the spill's location. Removing 5 to 10 per cent of an oil spill is about the best crews can achieve.

Prevention is the best tactic. Coastal jurisdictions are requiring oil tankers to abide by regulations that govern their mechanical and electrical maintenance. Some oil companies are also setting their own standards and refusing to use ships that fail to meet them. The United States has promised that all of its tankers will have double hulls by the year 2010. Other prevention efforts include using escorts to guide oil tankers into inland waters and setting speed limits in crowded or perilous waters.

Whatever efforts are taken to prevent oil spills—or any form of sea pollution—require the support of an informed, concerned public. Along the emerald sea, there are several research and educational facilities geared to interest people in the rich sea life off their coasts and make them aware of the need to protect it. The Oregon Coast Aquarium opened in Newport in May 1992 with indoor and outdoor exhibits of more than 150 animal species. Nearby is another large public display at the Oregon State University

A red Irish lord rests on a bed of plumose anemones. Colourful red Irish lords are most often found in areas of strong currents.

Exploring a Shipwreck

*Approaching the wreck of the **Husky**, a small commercial fishing boat sunk in the clear waters of Jervis Inlet, we surprise a harbour seal hiding in the hold. It looks out at us in astonishment, then bolts away. The **Husky** is typical of many small wrecks in the emerald sea. It has only been down for a few years but already it is decorated with colourful sea anemones and is home to a wide variety of sea life. Quillback rockfish hide in its many nooks and crannies, while exquisite nudibranchs crawl slowly along its hull. Feather stars perch on some of the deck fittings. Shipworms are already at work on the wooden hull, and in a few more years all that will remain of the **Husky** will be the rusted hulk of its machinery.*

Mark O. Hatfield Marine Science Center. The Vancouver Public Aquarium in British Columbia also features a broad cross-section of the species that live in the emerald sea.

While the numbers of recreational divers are growing quickly—including a burgeoning contingent who suffer disabilities, such as muscular dystrophy and paraplegia—technology is increasingly making the underwater realm accessible to non-divers. Using a two-person recreational submarine, a licensed pilot can fly a passenger through the water farther and faster than a scuba diver can travel. The cockpit of the tiny sub traps a pocket of air, which allows normal breathing and talking. Even more user-friendly are tourist submarines—built on this coast—that take upwards of twenty-eight passengers plus crew in buslike comfort to view the undersea world through large acrylic windows. These larger subs operate in the tropics, where the year-round tourist season makes them commercially viable.

The more people explore the sea and learn about its diverse life, the more likely they are to support efforts to establish marine sanctuaries. There are few areas in the emerald sea where consumptive uses of sea life are completely disallowed. Although British Columbia and the coastal states have designated marine parks and reserves of various kinds, the welfare of sea life below the low-tide mark is a matter of federal jurisdiction.

A few sites have gained ongoing protection because they have exceptional value to researchers or because they represent a particularly unique or rich habitat; and the waters that humpback whales and other sea life frequent within Alaska's Glacier Bay National Park are protected because of the area's status as a terrestrial park. Beyond those sites, however, if protection exists at all, it is mainly ad hoc and often seasonal, taking the form of a fishing closure, for example.

Even British Columbia's Ecological Reserve designation,

introduced to conserve life or sites that are exceptional or particularly representative, offers limited protection to marine sites. Although the federal government has agreed to fishing closures among the extremely rich sea life communities at the Race Rocks reserve, for instance, it has not closed a critical killer whale habitat at the Robson Bight (Michael Bigg) reserve off Vancouver Island because it is a prime commercial fishing ground.

Up and down the coast, a growing movement is calling for stronger, long-lasting protection for sea life of all kinds. It is promoting the establishment of a system of marine sanctuaries to recognize areas worthy of conserving and maintain them in their natural state—in perpetuity. Such sanctuaries would offer recreational and educational opportunities for viewing a wide variety of sea life and protect rare and endangered species and habitats. They would provide researchers with baseline information and sites for conducting long-term investigations and allow resource managers the chance to test a variety of management

The *Emerald Maiden*

Beneath the dark waters a mermaid's arms reach up to us imploringly as we descend into her undersea kingdom. I slowly spiral down towards her. Several quillback rockfish hover in her shadow, while elegant stalks of plumose anemones stand like columns on the rocks below. I can see the great bronzed tail curled beneath her, as the mermaid stares silently towards the surface world. Although her face shows clearly the etchings of time beneath the sea, her beauty is as unmistakable and timeless as the sea itself.

The bronze statue called the **Emerald Maiden** stands on the sea floor in twenty metres of water, at Saltery Bay, British Columbia. The work of sculptor Simon Morris, it is a tribute to those who know and love the wonders of the emerald sea. But it will take more than admiration to preserve these natural wonders for future generations.

In **The Ocean World**, Jacques Cousteau relates that when he first went to sea on board the **Calypso**, he was "...convinced that the oceans were immense, teeming with life, rich in resources of all kinds. ...But soon I had to face the evidence: the blue waters of the open sea appeared to be, most of the time, a discouraging desert. Like the deserts on land, it was far from dead, but the live ingredient, plankton, was thinly spread, like haze, barely visible and monotonous. Then, exceptionally, areas turned into meeting places; close to shores and reefs, around floating weeds or wrecks, fish would gather and make a spectacular display of vitality and beauty. Years of diving have revealed to me that the same situation occurs on the bottom of the sea. On the floor as in midwater, endless deserts are spotted with rare but exuberant oases."

We have the privilege of living on the shores of one of these rare "oases," but the ocean deserts exist here as well. Even beneath the emerald sea, the truly remarkable areas, awe-inspiring in their richness and vitality, are few and far between. These spectacular underwater wilderness areas deserve the same respect and protection that we would give to their counterparts on land. The future of the emerald sea is in our hands.

To escape danger, a giant nudibranch is able to swim slowly by twisting and turning its body gracefully through the water.

techniques. And they would also function as refuges and natural nurseries that, among other things, would increase the productivity of commercial species.

In the United States, conservation organizations, such as the Center for Marine Conservation, advocate marine sanctuaries in Puget Sound and along the outer coast. In Canada, the Marine Life Sanctuaries Society of British Columbia works with local-interest groups to identify areas worth preserving, such as a passage off Gabriola Island in the Strait of Georgia and four spectacular sites in the Queen Charlotte Strait near Vancouver Island: Hunt Rock, Barry's Islet, Browning Pass, and Stubbs Island (now designated a marine park). The society is also encouraging the preservation of a few large areas, such as the west coast of the Queen Charlotte Islands, which are still relatively pristine.

Some divers are guilty of taking advantage of the lack of regulations protecting sea life, recklessly denuding sites or robbing them of food species. Many more, however, support the concept of marine sanctuaries and donate time, energy, and enthusiasm to helping establish a system of reserves.

Diving communities also work to protect artificial reefs—in the form of sunken ships—as habitat for sea life. Along the coast from Oregon to Southeastern Alaska, popularly called the "graveyard of the Pacific," the sea is strewn with wrecked ships of almost every kind. For reasons as various as storms, rough seas, hidden rocks, navigation errors, and equipment failures, these ships

A sea lemon nudibranch roams across a submerged dock piling.

died in the sea, but their remains gained new life as they became repositories for life-sustaining nutrients and shelter for a variety of plants and animals.

Just off Savary Island in the Strait of Georgia, for instance, the 142-tonne, Canadian cargo steamer SS *Capilano I* sank in 1915 when it struck a rock. It came to rest on its keel in thirty-seven metres of water where it managed to escape severe beatings from surface waves and strong currents. There, on a barren sandy floor, the ship has become an oasis in an undersea desert, starkly outlined by rows of white plumose anemones. The bow is home to octopuses, Puget Sound king crabs, and several species of rockfish, including tiger, quillback, and copper rockfish. The roofless shell of what was once the wheelhouse is sheathed in sea anemones, sponges, barnacles, and crabs. Lingcod frequently occupy the stern and the hold, and the nearly two-metre-tall propeller is buried in anemones.

Working with preservation and research groups, such as the Underwater Archaeological Society of British Columbia and the Argonaut Society of the Puget Sound region, divers promote an ethic of conservation for wrecks, such as the *Capilano I*, which has been designated an underwater heritage site. Another such wreck is the fifty-eight-metre sidewheel passenger steamer SS *Del Norte,* an 1868 casualty of flood tides and rocks in the Gulf Island's Porlier Pass. Although there is little left of this once-elegant ship, it is a haven for basket and feather stars, large sea cucumbers, coral, schools of black rockfish, lingcod, kelp greenlings, and seaperch.

In Washington waters, the remains of the American freighter MS *Diamond Knot* have formed a fascinating reef. Ploughing through the foggy Strait of Juan de Fuca in 1947, a much larger freighter struck the 5,012-tonne, Seattle-bound *Knot*, which sunk to a depth of forty metres near Crescent Bay. In recovering 2 million dollars of cargo—mainly choice canned salmon—salvagers cut away sections of the ship's metal hull. Today, that hull houses some of the region's biggest wolf-eels, and the *Knot* is also habitat for lingcod, quillback and yelloweye rockfish, and groves of white and orange plumose anemones.

More recently, the 7,800-tonne freighter *Vanlene* ran aground in 1972 near Austin Island in Barkley Sound while trying to transport three hundred Dodge Colts from Japan to Vancouver. The sea continues to pound the *Vanlene*, strewing sections of the ship and its cargo about, but lingcod have moved into the wreck's many nooks and crannies and plumose anemones blanket the edges of steel plates. Thousands of rockfish school among the remains, where they rest at night.

Not all life-supporting artificial reefs are the results of accidents; some inoperative ships are deliberately scuttled to provide habitat for sea life. The Artificial Reef Society of British Columbia encourages the use of these ships to help compensate for the loss of some of the region's natural reefs to industry or development.

The society prepared two ships for sinking by stripping off materials that might be hazardous to sea life and making the ves-

Exploring Discovery Passage off northern Vancouver Island, a diver observes several quillback rockfish.

Opposite: A crescent gunnel peers out from a small hole in a colourful rock face. These eel-shaped fish are secretive and can often be found under rocks, in crevices, or hiding among the debris around wharves.

sels safer for divers by removing doors and cutting hatches. In 1991, it sank the fifty-four-metre freighter *GB Church* in Princess Margaret Marine Park near Sidney. The ship settled squarely on its keel and quickly gained tenants, such as octopuses, wolf-eels, rockfish, lingcod, nudibranchs, sea anemones, crabs, and tube worms. The following year, the society scuttled the 112-metre naval destroyer escort HMCS *Chaudiere* in Sechelt Inlet, creating the largest artificial reef on the west coast of North America and the third-largest in the world. Although the *Chaudiere* did not land as planned, sinking on its side, it will accommodate increasing numbers of invertebrates, fish, and other sea life through the years ahead.

* * *

By developing and adopting a sound ethic of conservation, we can hope to protect and enhance this special sea and the many species with whom we share it. We can also hope to protect ourselves. From deep-diving mammals, we are learning about human heart attacks. From disease-resistant sharks, we are learning about illness prevention in people. From king crabs, we are deriving a substance to prevent the spread of the human immunodeficiency virus (HIV) within the body. But even more important, as ocean explorer Jean-Michel Cousteau insists, each species acts as a tiny shock absorber for the whole system, helping the human race and nature itself recover when catastrophe hits.

On a visit to the emerald sea in 1992, Cousteau stressed that society should not look to technology to save the sea, but rather it should look to itself. "We are going to succeed from the heart, not the mind," he said.

There is no better time to act from the heart.

Scientific Names of Species

Common names of species vary according to region and time. Scientific names change as species are further studied and reclassified. This list includes names in use at the time of printing.

Aggregating anemone *Anthopleura elegantissima*
Alabaster nudibranch *Dirona albolineata*
Basking shark *Cetorhinurs maximus*
Bay pipefish *Syngnathus griseolineatus*
Big skate *Raja binoculata*
Black-clawed crab *Lophopanopeus bellus*
Black rockfish *Sebastes melanops*
Blood star *Henricia leviuscula*
Blue-clawed lithode crab *Oedignathus inermis*
Blue mussel *Mytilus edulis*
Blue shark *Prionace glauca*
Blue whale *Balaenoptera musculus*
Boot sponge *Rhabdocalyptus dawsoni*
Boring sponge *Cliona celata*
Box crab *Lopholithodes foraminatus*
Brooding anemone *Epiactis prolifera*
Bull kelp *Nereocystis luetkeana*
Butter clam *Saxidomus giganteus*
Cabezon *Scorpaenichthys marmoratus*
Calcareous tube worm *Serpula vermicularis*
California blue mussel *Mytilus californianus*
California sea cucumber *Parastichopus californicus*
California sea lion *Zalophus californianus*
Candy stripe shrimp *Lebbeus grandimanus*
China rockfish *Sebastes nebulosus*
Clinging jellyfish *Gonionemus vertens*
Cloud sponge *Aphrocallistes vastus* and *Chonelasma calyx*
Clown shrimp *See* Candy stripe shrimp
Club-tipped anemone *See* Strawberry anemone
Copper rockfish *Sebastes caurinus*
C-O sole *Pleuronichthys coenosus*
Crescent gunnel *Pholis laeta*
Crimson anemone *Cribrinopsis fernaldi*
Dahlia anemone *Urticina crassicornis*
Dall's porpoise *Phocoenoides dalli*
Decorated warbonnet *Chirolophis decoratus*
Decorator crab *Oregonia gracilis*
Deep-water finger sponge *Neoespeiropsis digitata*
Deep-water gorgonian coral *Paragorgia pacifica*
Dentalium *Dentalium pretiosum*
Dungeness crab *Cancer magister*
Elephant seal *Mirounga angustirostris*
Encrusting sponge *Haliclona permollis*
False killer whale *Pseudorca crassidens*
Feather-duster tube worm *Eudistylia vancouveri*
Fish-eating anemone *Urticina piscivora*
Giant barnacle *Balanus nubilus*
Giant green anemone *Anthopleura xanthogrammica*
Giant nudibranch *Dendronotus iris*
Giant Pacific octopus *Octopus dofleini*
Gooseneck barnacle *Mitella polymerus*
Gorgonian coral *Calcigorgia spicculifera*
Green sea urchin *Strongylocentrotus droebachiensis*
Grey whale *Eschrichtius robustus*
Grunt sculpin *Rhamphocottus richardsoni*
Hairy hermit crab *Pagurus hirsutiusculus*
Harbour porpoise *Phocoena phocoena*
Harbour seal *Phoca vitulina*
Heart crab *Phyllolithodes papillosus*
Hooded nudibranch *Melibe leonina*
Humpback whale *Megaptera novaeangliae*
Japanese oyster *Crassostrea gigas*

Japanese weed *Sargassum muticum*
Kelp greenling *Hexagrammos decagrammus*
Kelp perch *Brachyistius frenatus*
Killer whale *Orcinus orca*
Leather star *Dermasterias imbricata*
Lingcod *Ophiodon elongatus*
Lion nudibranch *See* Hooded nudibranch
Long ray sea star *Stylasterias forreri*
Masking crab *Scyra acutifrons*
Minke whale *Balenoptera acutorostrata*
Moon jelly (jellyfish) *Aurelia aurita*
Mosshead warbonnet *Chirolophis nugator*
Northern abalone *Haliotis kamtschatkana*
Northern fur seal *Callorhinus ursinus*
Northern kelp crab *Pugettia producta*
Ocean sunfish *Mola mola*
Ochre sea star *Pisaster ochraceus*
Opalescent nudibranch *Hermissenda crassicornis*
Orange cup coral *Balanophyllia elegans*
Orange decorator crab *Chorilia longipes*
Orange finger sponge *Neoespeiropsis rigida*
Orange hermit crab *Elassochirus gilli*
Orange-peel nudibranch *Tochuina tetraquetra*
Orange sea cucumber *Cucumaria miniata*
Orange sea pen *Ptilosarcus gurneyi*
Orange-spotted nudibranch *Triophia catalinea*
Orca *See* Killer whale
Oregon triton *Fusitriton oregonensis*
Pacific herring *Clupea pallasi*
Pacific sandlance *Ammodytes hexapterus*
Pacific white-sided dolphin *Lagenorhynchus obliquidens*
Painted anemone *See* Dahlia anemone
Painted greenling *Oxylebius pictus*
Perennial kelp *Macrocystis integrifolia*
Pink soft coral *Gersemia rubiformis*
Plume worm *See* Calcareous tube worm
Puget Sound king crab *Lopholithodes mandtii*
Purple sea star *See* Ochre sea star
Purple sea urchin *Strongylocentrotus purpuratus*
Quillback rockfish *Sebastes maliger*
Ratfish *Hydrolagus colliei*
Red-gilled aeolid nudibranch *Flabellina salmonacea*
Red Irish lord *Hemilepidotus hemilepidotus*
Red sea urchin *Strongylocentrotus franciscanus*
Ringed top snail *Calliostoma annulatum*
Sailfin sculpin *Nautichthys oculofasciatus*
Salmon-gilled aeolid nudibranch *Flabellina verrucosa* and *Flabellina fusca*
Sand dollar *Dendraster excentricus*
Scalyhead sculpin *Artedius harringtoni*
Sea blubber *Cyanea capillata*
Sea lemon *Archidoris montereyensis*
Sea otter *Enhydra lutris*
Sea palm *Postelsia palmaeformis*
Shipworm *Bankia setacea*
Silver surfperch *Hyperprosopon ellipticum*
Sixgill shark *Hexanchus griseus*
Sockeye salmon *Oncorhynchus nerka*
Soupfin shark *Galeorhinus Zyopterus*
Sperm whale *Physeter macrocephalus*
Spiny dogfish shark *Squalus acanthias*
Spiny pink scallop *Chlamys hastata*
Staghorn bryozoan *Heteropora pacifica*
Steller's sea lion *Eumetopias jubatus*
Strawberry anemone *Corynactis californica*
Striped seaperch *Embiotoca lateralis*
Sulphur sponge *Verongia thiona*
Sunflower star *Pycnopodia helianthoides*
Swimming anemone *Stomphia didemon*
Swimming scallop *See* Spiny pink scallop
Three-coloured polycera nudibranch *Polycera tricolor*
Tidepool sculpin *Oligocottus maculosus*
Tiger rockfish *Sebastes nigrocinctus*
Tube-dwelling anemone *Pachycerianthus fimbriatus*
Vermilion star *Mediaster aequalis*
Water jellyfish *Aequorea victoria*
White shark *Carcharodon carcharias*
White-spotted anemone *Urticina lofotensis*
Wolf-eel *Anarrhichthys ocellatus*
Yelloweye rockfish *Sebastes ruberrimus*
Yellowtail rockfish *Sebastes flavidus*

Suggested Reading

Bodsworth, Fred. *The Pacific Coast*. Toronto: Natural Science of Canada Limited, 1970.

Buchsbaum, Ralph et al. *Animals Without Backbones*, Third Edition. Chicago: The University of Chicago Press, 1987.

Bulloch, David K. *The Underwater Naturalist: A Laymen's Guide to the Vibrant World Beneath the Sea*. New York: Lyons & Burford, 1991.

Carefoot, Thomas. *Pacific Seashores: A Guide to Intertidal Ecology*. Vancouver: J.J. Douglas Ltd., 1977.

Carl, G. Clifford. *Guide to Marine Life of British Columbia* (Handbook No. 21). Victoria: British Columbia Provincial Museum, 1971.

Cousteau, Jacques Yves. *Jacques Cousteau: The Ocean World*. New York: Harry N. Abrams, Inc., 1985.

Cribb, James. *Treasures of the Sea: Marine Life of the Pacific Northwest*. Toronto: Oxford University Press, 1983.

Downing, John. *The Coast of Puget Sound: Its Processes and Development*. Seattle: University of Washington, 1983.

Drew, Wayland et al. *The Nature of Fish*. Toronto: Natural Science of Canada Limited, 1974.

Ellis, Richard. *The Book of Sharks*. New York: Alfred A. Knopf, 1989.

Gibbs, James A. *Shipwrecks of the Pacific Coast*, Second Edition. Portland: Binfords & Mort, 1962.

Harbo, Rick M. *Guide to the Western Seashore: Introductory Marinelife Guide to the Pacific Coast*. Surrey: Hancock House Publishers Ltd., 1988.

——*Tidepool and Reef: Marinelife Guide to the Pacific Northwest*. Surrey: Hancock House Publishers Ltd., 1984.

Herger, Bob and Neering, Rosemary. *The Coast of British Columbia*. Vancouver: Whitecap Books Ltd., 1989.

Hoyt, Erich. *The Whales of Canada*. Camden East: Camden House Publishing Ltd., 1984.

King, Judith E. *Seals of the World*, Second Edition. London: British Museum (Natural History), 1983.

Kozloff, Eugene N. *Seashore Life of Puget Sound, the Strait of Georgia, and the San Juan Archipelago*. Vancouver: J.J. Douglas Ltd., 1973.

Lamb, Andy and Edgell, Phil. *Coastal Fishes of the Pacific Northwest*. Madeira Park: Harbour Publishing Co. Ltd., 1986.

Lambert, David and McConnell, Anita. *Seas and Oceans*. New York: Facts on File, 1987.

Leatherwood, Stephen and Reeves, Randall R. *Handbook of Whales and Dolphins*. San Francisco: Sierra Club Books, 1983.

Leon, Vicki. *A Raft of Sea Otters: An Affectionate Portrait*. San Luis Obispo: Blake Publishing, 1988.

——*Seals and Sea Lions*. San Luis Obispo: Blake Publishing, Inc., 1988.

Liburdi, Joe and Truitt, Harry. *A Guide to Our Underwater World*. Saanichton: Hancock House Publishers, 1973.

Obee, Bruce and Ellis, Graeme. *Guardians of the Whales: The Quest to Study Whales in the Wild*. Vancouver: Whitecap Books Ltd., 1992.

Obee, Bruce and Fitzharris, Tim. *Coastal Wildlife of British Columbia*. Vancouver: Whitecap Books Ltd., 1991.

Paine, Stefani Hewlett. *Beachwalker: Sea Life of the West Coast*. Vancouver: Douglas & McIntyre, 1992.

Parker, Henry S. *Exploring the Oceans: An Introduction for the Traveler and Amateur Naturalist*. Englewood Cliffs: Prentice Hall, Inc., 1985.

Paterson, T.W. *British Columbia Shipwrecks*. Langley: Stagecoach Publishing Co. Ltd., 1976.

Pratt-Johnson, Betty. *141 Dives in the Protected Waters of Washington and British Columbia*, Updated Edition. Vancouver: Gordon Soules, 1977.

Rogers, Fred. *Shipwrecks of British Columbia*. Vancouver: J.J. Douglas Ltd., 1973.

——*More Shipwrecks of British Columbia*. Vancouver: Douglas & McIntyre, 1992.

Rotman, Jeffrey L. and Allen, Barry W. *Beneath Cold Seas: Exploring Cold Temperate Waters of North America*. New York: Van Nostrand Reinhold Co., 1983.

Sackett, Russell et al. *Edge of the Sea*, Revised Edition. Alexandria: Time-Life Books, 1985.

Sargent, William. *The Year of the Crab: Marine Animals in Modern Medicine*. New York: W.W. Norton & Co., 1987.

Snively, Gloria. *Exploring the Seashore in British Columbia, Washington and Oregon: A Guide to Shorebirds and Intertidal Plants and Animals*. Vancouver: Gordon Soules Book Publishers Ltd., 1978.

Thomson, Richard, E. *Oceanography of the British Columbia Coast*. Ottawa: Department of Fisheries and Oceans, 1981.

Waaland, J. Robert. *Common Seaweeds of the Pacific Coast*. Vancouver: J.J. Douglas Ltd., 1977.

Whipple, A.B.C. et al. *Restless Oceans*, Revised Edition. Alexandria: Time-Life Books, 1984.

Periodicals

DIVER Magazine. Richmond: Seagraphic Publications Ltd.

Rodale's *Scuba Diving*. Savannah: Rodale Press, Inc.

Converting Metric Measurements

Here is a simple table to help you understand the metric system.

Fahrenheit/Celcius

When you know	Multiply by	To find
millimetres	.04	inches
centimetres	.4	inches
metres	3.3	feet
kilometres	.63	miles
litres	.26	gallons
grams	.04	ounces
kilograms	2.2	pounds
tonnes	1.1	tons

or:

inches	.25	millimetres
inches	2.5	centimetres
feet	.3	metres
miles	1.6	kilometres
gallons	3.8	litres
ounces	28	grams
pounds	.45	kilograms
tons	.9	tonnes

Temperatures are given in Celsius and their relationship to the Fahrenheit scale is shown at right. To convert temperatures in Celsius to Fahrenheit, multiply by 9/5, then add 32.

Index

Photographs are indicated in **bold**.

Abalone, 22, 27, 36-37, 65
  See also specific names
Aggregating anemones, 46
Alabaster nudibranchs, 66, **77**, 77
Alaska Range, 7
Algae, 17, 31, 36, 37, 56, 57, 58, 60, 66, 71
  See also specific names
Anacortes, 126, 128
Anemones
  See Sea anemones
Argonaut Society, 134
Arran Rapids, 50
Artificial reefs, x, 126, 133-136
Artificial Reef Society of British Columbia, 135
Atlantic coast, 7, 8, 11, 114
Auklets, 26
Austin Island, 135

Baines Bay, 49
Baranof Island, 33
Barkley Sound, 34, 98, 99, 103, 111, 118, 135
Barnacles, 22, 27, 35-36, 46, 49, 51, 63, 86, 103, 106, 123, 127, 134
  See also specific names
Barracuda, 11
Barry's Islet, 133
Basket stars, **50**, 51, **80**, 81, 134
Basking sharks, 2, 95, 98, 101, 118
Bay pipefish, 91, 93
Becher Bay, 114-115
Bellingham, 126
Big skates, 94
Bioluminescence, 61, 63
Black-clawed crabs, 78
Black rockfish, 18-19, 134
Blood stars, **3**, 81
Blue-clawed lithode crabs, 50
Blue mussels, 49
Blue sharks, 10
Blue whales, 41, 63
Bonito, 11
Boot sponges, 17
Boring sponges, 69
Botanical Beach, 33
Botany Bay, 33
Box crabs, 78
Brooding anemones, **1**, 1, 2, 31, 46, 49, 57
Browning Pass, **43**, 50-51, **118**, 118, 133
Bryozoans, 22, **77**
  See also specific names
Bull kelp, 23, 27, 36, **53**, 54, 55-57, **56**, 58, **58-59**
Bute Inlet, 50
Butter clams, 63

Cabezons, 55, 87, **99**, 99
Calcareous tube worms, **6**, 49
California blue mussels, 35
California sea cucumbers, 82
California sea lions, 23, 109
Candy stripe shrimp, 21, 29, 51, **52**, 74
Cape Flattery, 33, 126
Cascade Mountains, 21
Center for Marine Conservation, 133
Chatham Strait, 7, 14
Cherry Point, 128
Chichagof Island, 33
China rockfish, 37-38, **40**, 89, 90, **100**, 101
Chitons, 22
Circulation, water, 11, 15, 19, 27, 45
Clams, 63, 65, 81, 107, 127
  See also specific names
Clinging jellyfish, 57
Cloud sponges, 2, **13**, 13, **16**, 17-18, 27, 70
Clown shrimp
  See Candy stripe shrimp
Club-tipped anemones
  See Strawberry anemones
Cnidarians, 70-77
  See also specific names
Coastal Trough, 7
Continental shelf, 7-8, 10-11, 29
Copper rockfish, 18, 55, 134
Coralline algae, **24-25**, 36, 57
  See also specific names
Corals, 2, **40**, 41, 50, 70, 74, 77, 89, 134
  See also specific names
C-O soles, **95**, 95
Cousteau, Jacques, 132
Cousteau, Jean-Michel, 136
Crabs, 17, 21, 22, 27, 35, 56, 63, 64, 78-79, 86, 91, 98, 127, 134, 136
  See also specific names
Crescent Bay, 135
Crescent gunnels, 136, **137**
Crimson anemones, **5**, 6, 21, 29, 51, **52**, **72-73**, 74, **76**, **119**
Crinoids
  See Feather stars
Crustaceans, 77-79
  See also specific names
Currents, ix, 1, 10, 11, 15, 21, 27, 30, 34, 41, 42, 45-47, 49-51, 59, 126, 134
  See also specific names

Dahlia anemones, 46, 71, 74, **78**, 78
Dall's porpoises, 103
Decorated warbonnets, 46, 49, **92**, 93
Decorator crabs, 17, 18, 41
Deep-water gorgonian corals, **14**, 14, 18, 123

Denali, 7
Dentalia, 115-116
Dinoflagellates, 61, 63
Discovery Passage, 21, 46-47, 135
Dogfish
  See Spiny dogfish sharks
Dolphins, 103
  See also specific names
Dungeness crabs, 78

Echinoderms, 79-82
  See also specific names
Ecological Reserves, 130-131
Edmonds, 22
Eelgrass, 17, 57, 66, 78, 91, 93, 128
Elephant seals, 111, 116
El Nino, 10-11
Emerald Maiden, **132**, 132
Encrusting sponges, ix, 51
English Bay, 103
Exxon Valdez, xi, 127

False killer whales, 103
Feather-duster tube worms, **36**, 36, 47
Feather stars, 2, **12**, 12, **15**, 17, 130, 134
Finger sponges, **43**, 46, 51, **118**
Finlayson Channel, 14
Fiords, xi, 1, 7, 11-18, 27
  See also specific names
Fish, 2, 11, 17, 21, 27, 37, 54, 55, 57, 70, 81, 84-101, 105, 109, 118-119, 123, 124, 126, 132, 136
  See also specific names
Fish-eating anemones, 35, 37, 38, **38-39**, 74, **83**, 83
Fishing, 2, 115, 116, 118-119, 130, 131
Flora Islet, 98
Flounders, 93-94
Fraser River, 21, 26
Fur seals
  See Northern fur seals

Gabriola Island, 133
Galiano Island, 49
GB Church, 136
Giant barnacles, **xii**, 2, 50
Giant green anemones, **31**, 31, 37, 71
Giant kelp, 36
Giant nudibranchs, 29, 66, **68**, 68, 69, **133**, 133
Giant Pacific octopuses, **26**, 26, 27, 49, 54, 63-65, **64**, **65**, 74, 75
Glaciation, 7, 11, 14-15, 19
Glacier Bay, 14
Glacier Bay National Park, 130
Gooseneck barnacles, 35, 49-50, **81**, 81
Gorgonian corals, x, **46**, 49, 49, 51, 77
  See also Deep-water gorgonian corals

Grays Harbor, 126
Greenlings, 91
    See also specific names
Green sea urchins, 82
Grey whales, x, 39, 41, 58, **102**, 103, 105-106, 118
Grunt sculpins, 27, **87**, 87
Gulf Islands, **18**, 19, 21, 22, 26-27, 42, 134
    See also specific names

Hairy hermit crabs, 79
Harbour porpoises, 103
Harbour seals, ix, 23, **107**, 107, 111, 116, 130
Harvesting, xi, 2, 115-116, 122-123, 124, 127, 133
Heart crabs, 79, **83**, 83
Hecate Strait, 7
Hermit crabs, **29**, 29, 71, 79
    See also specific names
Herring, 2, 118
    See also specific names
HMCS *Chaudiere*, x, **115**, 126, **127**, 136
HMCS *Thiepval*, 34-35
Hooded nudibranchs, 66, **79**, 79
Hornby Island, 98
Humpback whales, 2, 105, 106-107, 118, 130
Hunt Rock, 89, 133
Husky, 130, **131**
Hydrocorals, **3**, 3, **62**, 63, **71**
Hydroids, **xii**, xii, **37**, **47**, 51, 116

Inner coasts, xi, 2, 11, 18-29, 41, 59-60, 103, 106
Inside Passage, 19
Invertebrates, 2, 51, 54, 63-83, 105, 109, 119, 122-123, 136
    See also specific names

Japanese Current, 8
Japanese oysters, 60
Japanese weed, 60
Jellyfish, 12, 27, 70, 77
    See also specific names
Jervis Inlet, **11**, 11, 15, 130
Johnstone Strait, 19

Kelp, ix, 2, 17, 18, 22, 31, 41, 42, 57, 109, 110
    See also specific names
Kelp greenlings, **20**, 27, 35, 55, **90**, 90, **125**, 134
Kelp perch, 55
Killer whales, 22-23, 41, 103-105, **104-105**, 106, 116, 127, 131
Krill, 106
Kuroshio Current, 8
Kyuquot Sound, 116

Lark, 33
Leather stars, 71
Limpets, 35, 127
Lingcods, **33**, 33, **34**, 35, 91, 134, 135, 136
Lion nudibranchs
    See Hooded nudibranchs

Long ray sea stars, 81

Mammals, 2, 54, 101-111, 116, 119, 123, 136
    See also specific names
Marine Life Sanctuaries Society of British Columbia, 133
Marine parks, xi, 130, 133
Marine sanctuaries, 130-131, 133
Masking crabs, **19**, 19
Minke whales, 26, 105, 106
Molluscs, 63-69
    See also specific names
Moon jellies (jellyfish), **8**, 8, 77
Morris, Simon, 132
Mosshead warbonnets, 18, 93
Mount Logan, 7
Mount McKinley
    See Denali
MS *Diamond Knot*, 135
Mussels, 2, 27, 35-36, 49, 81, 123, 124, 127
    See also specific names

Nakwakto Rapids, 42, 49-50, 81
Nestucca, 126
Newport, 128
Newtsuits, 123
Northern abalone, 37
Northern fur seals, 41, 109-111, 116
Northern kelp crabs, 56
North Pacific Drift, 8
Nudibranchs, 22, 41, 65-69, **66-67**, 70, 74, 130, 136
    See also specific names
Nutrients, ix, 10, 11, 17, 27, 35, 45, 54, 57, 59, 63, 134

Ocean sunfish, 10
Ochre sea stars, 8, **9**, 17, 49
Octopuses, ix, **xi**, 2, 18, 122, 134, 136
    See also specific names
Ogden Point Breakwater, 22
Oil spills, xi, 126-128
Olympia, 19
Olympic Peninsula, 19, 124
Opalescent nudibranchs, 17, 23, **24-25**
Orange cup corals, **x**, **60**
Orange decorator crabs, **76**
Orange hermit crabs, 79
Orange-peel nudibranchs, ix, **50**, 50, 51
Orange sea cucumbers, 82
Orange sea pens, **20**, 21, 22, 61
Orange-spotted nudibranchs, 56, **60**
Orcas
    See Killer whales
Oregon Coast Aquarium, 128
Oregon State University Mark O. Hatfield Marine Science Center, 128, 130
Oregon tritons, **29**
Outer coasts, xi, 2, 11, 23, 27, 29-41, **30**, 46, 47, 49, 50, 59, 103, 106, 124, 133
Oxygen in water, 17, 21, 35, 45, 63
Oysters, 63, 65, 112, 127
    See also specific names

Pacific herring, 23, 26, **56**, 57

Pacific salmon, 26, 87
    See also specific names
Pacific sandlances, **56**, 57
Pacific white-sided dolphins, 101-103, **108**, 109
Painted anemones
    See Dahlia anemones
Painted greenlings, **91**, 91
Perch, 22
    See also specific names
Perennial kelp, 36
Phytoplankton, 1, 10, 11, 53, 60, 63
Pink soft corals, ix, **xi**, **43**, 50-51, **72-73**, 77, **114**, 114, **118**
Pinnipeds, 109-111
    See also Seals, Sea lions
Pipefish
    See Bay pipefish
Plankton, 13, 14, 22, 50, 54, 63, 98, 105, 114, 118, 132
Plume worms
    See Calcareous tube worms
Plumose anemones, **xii**, 18, 21, 22, 31, 37, **47**, 49, 71, **82**, 82, **93**, **112**, **117**, **128-129**, 132, 134, 135
Pollution, xi, 2, 124, 126-128
    See also specific kinds
Porifera, 69-70
    See also specific names
Porlier Pass, 49, 134
Porphyra nereocystis, 56
Porpoises, 103
    See also specific names
Port Angeles, 126
Powell Lake, 15
Prince of Wales Island, 33
Princess Margaret Marine Park, 136
Prince William Sound, xi, 127
Puget Sound, 5, 7, 18-29, 103, 111, 122, 124, 133
Puget Sound king crabs, 18, 35, 74, **75**, 78-79, 134
Pulp mill effluent, 124
Purple sea stars
    See Ochre sea stars
Purple sea urchins, 37, 41

Quadra Island, 21
Queen Charlotte Islands, 7, 8, 10, 13, 33, 36, 105, 133
Queen Charlotte Sound, 7
Queen Charlotte Strait, x, xi, 19, 42, 49, 50-51, 89, 109, 118, 133
Quillback rockfish, **10**, 18, 27, 130, 132, 134, **135**, 135

Race Rocks, 131
Ratfish, **94**, 94
Red-gilled aeolid nudibranchs, **x**
Red Irish lords, ix, **44-45**, 45, 47, **54**, **84-85**, 86, 87, 88, **96-97**, **128-129**, 129
Red sea urchins, **4**, 21, **23**, 23, **32**, 57, **83**, 109
Red tides, 63
Remotely operated vehicles (ROVs), 123
Ringed top snails, **116**, 116
Ripple Rock, 42, 45
Robson Bight, 131

Rock crust, 36
Rockfish, ix, 7, 10, **13**, 17, 27, 30, 37, 41, 90-91, 94, 126, **134**, 135, 136
  See also specific names
Rockweed, **55**
Rough strap, 36

Sailfin sculpins, 89-90, **93**, 93
St. Elias Mountains, 7
Salinity, 15, 21, 45
Salish, 115
Salmon, 11, 115, 118
  See also specific names
Salmon-gilled aeolid nudibranchs, **47**
Saltery Bay, 132
Sand dollars, 79, 82
San Juan Islands, 19, 21, 26-27, 42, 66
Saranac, 42
Savary Island, 134
Scallops, 22, 69
  See also specific names
Scalyhead sculpins, **101**, 101
Sculpins, 87-90
  See also specific names
Sea anemones, ix, x, **xi**, 2, 22, 27, **28**, 31, 35, 37, **40**, 41, **43**, **44-45**, 45, 46, **48**, 51, 70-74, 89, 91, **96-97**, **118**, 126, 130, 134, 136
  See also specific names
Seabirds, 23, 57, 109, 126, 127
  See also specific names
Sea blubber, 77
Sea cucumbers, 22, 79, 82, **119**, 119, 134
  See also specific names
Sea grasses, 57, 60
  See also specific names
Sea lemons, **134**
Sea lettuce, 18, 58
Sea lions, x, 2, 19, 23, 26, 63, 95, 105, 109-111, 116
  See also specific names
Seals, 95, 105, 109-111, 116
  See also specific names
Sea otters, 2, **106**, 106, 107, 109, 116, 127
Sea palms, 2, 36
Sea pens, 18, 51, 63, 77
  See also specific names
Seaperch, 17, 21, 134
  See also specific names
Sea slugs
  See Nudibranchs
Sea stars, 2, 17, 21, **28**, **32**, 35-36, 49, 79, 81
  See also specific names
Seattle, 22
Sea turtles, 10
Sea urchins, 22, 27, 79, 81-82, 86, 89, **106**, 127
  See also specific names

Seaweed, 2, 17, 21, 30, 36, 41, 56, 57, 58, 60, 81
  See also specific names
Sechelt Inlet, 136
Sechelt Narrows (Rapids), **42**, 42, **44-45**, 45, 50
Sewage, 126
Seymour Narrows, 42, 50
Sharks, 95, 118-119, 136
  See also specific names
Ships, 14, 33, 34, 113, 135
  See also specific names
Shipworms, 130
Shipwrecks, x, 2, 27, 33, 34, 42, 45, **112**, 113, 130, 132, 133-135
  See also specific names
Shrimp, 86, 89
  See also specific names
Sidney, 136
Sills, 14-15
Silt, 7, 14, 15, 17, 21, 23, 45
Silver surfperch, 30
Sixgill sharks, 95, **98**, 98, 99, 119
Skates, 94
  See also specific names
Skookumchuck Rapids
  See Sechelt Rapids
Smithora naiadum, 57
Snails, 17, 35, 65, 81, 123-124, 127
  See also specific names
Sockeye salmon, 26, **113**, 114
Soupfin sharks, 10
Sperm whales, 41
Spiny dogfish sharks, **22**, 22, 26, 56, 95, 101
Spiny pink scallops, 49, **69**, 69
Sponges, **15**, 17, **19**, 22, **32**, **37**, **40**, 41, 47, 49, **69**, 69-70, 89, **90**, 134
  See also specific names
SS *Capilano I*, 134
SS *Del Norte*, 134
Staghorn bryozoans, 18, 31, **91**
Steller's sea lions, 23, **27**, 27, 38-39, **110**, 110
Strait of Georgia, 7, 15, **18**, 18-29, 49, 98, 111, 124, 133, 134
Strait of Juan de Fuca, 19, 49, 126, 127, 128, 135
Strawberry anemones, **10**, 46-47, 74, **86**, 86, **90**
Striped seaperch, 22, 116, **117**
Stubbs Island, 133
Submersibles, **123**, 123, 130
Sulphur sponges, **xi**, **43**, **118**
Sundancer, 45
Sunflower stars, **viii**, x, 17, **68**, 68, 69, 79
Sunlight in water, 13, 14, 55-56, 57, 58, 60, 63
Surfgrass, 30, 57

Swimming anemones, 71
Swimming scallops
  See Spiny pink scallops
Temperature, water, 8, 10, 11, 34, 45, 54, 109
Tenyo Maru, 127
Three-coloured polycera nudibranchs, **37**
Tidal passages, ix-xi, 2, 11, 27, 41-52
  See also specific names
Tidepool sculpins, 88
Tides, 3, 19, 41-42
  See also Red tides, Tidal passages
Tiger rockfish, **7**, 90, 119, **120-121**, **122**, 134
Tube-dwelling anemones, 66, 69, **70**, 70
Tube worms, 22, 46, 136
  See also specific names
Tyler Rock, 98

Underwater Archaeological Society of British Columbia, 134
Upwellings, 1, 10, 34, 59
Urchins
  See Sea urchins

Valdes Island, 49
Vancouver, 22, 103, 122, 123
Vancouver Island, x, 7, 8, 13, 18, 19, 21, 33, 34, 36, 42, 98, 102, 114, 116, 124, 126, 131, 133
Vancouver Public Aquarium, 130
Vanlene, 135
Vermilion stars, **4**
Victoria, 22, 126
Visibility, 13, 17, 21, 41

Wampum shells, 115-116
Warbonnets, 93
  See also specific names
Water jellyfish, **61**
Waves, 1, 5, 7, 11, 15, 21, 27, 29-30, **30**, 31, 33-37, 41, 45, 49, 50, 134
Whales, 26, 39, 54, 101-107, 109, 116, 118
  See also specific names
Whelks, 123-124
White sharks, 10, 95
White-spotted anemones, 74
Winds, 2, 8, 21, 29-30, 34, 126
Wolf-eels, 22, 86, **88**, **89**, 89, **122**, 122, 135, 136
Wrecks
  See Shipwrecks

Yelloweye rockfish, 135
Yellowtail rockfish, 91

Zoanthids, **32**, **119**